Jürgen M. Schneider

Protocol Engineering

Informatics

J. M. Schneider
Protocol Engineering
A Rule-based Approach

W. Bibel
Automated Theorem Proving

E. Eder
Relative Complexities of First Order Calculi

F. Kurfeß
Parallelism in Logic

Vieweg

Jürgen M. Schneider

Protocol Engineering

A Rule-based Approach

Die Deutsche Bibliothek - CIP-Einheitsaufnahme

Schneider, Jürgen M.:
Protocol engineering : a rule based approach / Jürgen M.
Schneider. - Braunschweig ; Wiesbaden : Vieweg, 1992

Verlag Vieweg, P.O. Box 5829, D-6200 Wiesbaden

Vieweg is a subsidiary company of the Bertelsmann Publishing International.

ISBN-13: 978-3-528-05243-0 e-ISBN-13: 978-3-322-83979-4
DOI: 10.1007/ 978-3-322-83979-4

Foreword

It is now over twenty years since the first efforts were made to interconnect computers for the exchange of data. In the 1970s proprietary networking architectures, such as IBM's SNA, were developed and installed at many customer sites worldwide. At the same time the ARPANET in the United States became a research vehicle providing many new insights, in particular into packet switching technology and routing algorithms.

In the 1980s the interconnection of heterogeneous equipment was a dominating issue: ISO published the Reference Model for Open Systems Interconnection (OSI) and many new standards for the protocols of layers 4 to 7. In parallel the INTERNET community developed a new generation of protocols, the TCP/IP protocol stack, in order to enable worldwide data interchange for scientists. Powerful workstations and PCs are now installed on many desktops because decentralized computing is often much more user friendly and cost effective than traditional mainframe computing. Yet access to shared information and resources is still required, creating a strong demand for interconnecting heterogeneous computers. Applications such as electronic mail, file transfer, client/server configurations and manufacturing automation are now in widespread use, with computer equipment from different vendors. The network has really become the computer.

In recent years the incredible advances in optical fiber technology have created a new physical medium for data transmission, with very high bandwidth and a very low bit error rate. The physical medium has increased in speed much faster than the computer. Originally most communication protocols were designed for relatively slow communication links and relatively fast CPUs. Thus the complete protocol stacks will now have to be revised for the new generation of high-speed networks. Due to the high data rates, these new networks will be able to support innovative applications. An example are interconnected multimedia workstations, integrating audio and video data streams with classical data communication. These advances in both networking technology and new applications generate new communications architectures and protocols at an increasing pace. In the future more and more protocols will have to be developed. The complex task of designing and implementing these protocols correctly requires the attitude of a good engineer: the use of a formal specification technique with well-defined syntax and semantics, a design methodology, and a toolkit to support the protocol designer.

This book presents a novel approach to protocol engineering. It is based on experience gained with earlier techniques, such as extended finite state machines, process algebra and Petri nets. The author gives a clear introduction into these techniques and illustrates their advantages and disadvantages. Then the new model of Communicating Rule Systems (CRS) is presented in detail, and operational semantics are defined. Using these operational semantics implementation concepts and runtime mechanisms are developed that facilitate the simulated execution of CRS specifications. The author describes the realization of these concepts, a program accomplishing the translation of a CRS specification into executable form, and an integrated toolkit for protocol engineering, including its implementation on a workstation. An outlook concludes the book. The connection establishment phase of the X.25 protocol is used as an example throughout the text, illustrating very clearly the various features of formal specification techniques and their implementation.

The book is an excellent introduction into the field of protocol engineering in general, and rule-based specification in particular. It also contains a comprehensive discussion of implementation issues for non-procedural specification languages which is hardly found anywhere else in the literature. The author works at IBM's European Networking Center in Heidelberg, and practical applicability of his ideas is as important to him as theoretical soundness. The book is intended for advanced students as well as for practitioners. To profit most from the text, the reader should have a basic knowledge of computer network architecture, and in particular of layered communication systems. Familiarity with predicate calculus or rule-based systems would also be helpful.

It is hoped that this book will contribute to an improved understanding of communication protocol development as an engineering task, based on formal models and supported by appropriate software toolkits.

Mannheim, March 1992 Wolfgang Effelsberg

Preface

Not long ago, the interconnection of computer systems was a highly complex task requiring considerable effort. Today, computer networks have become part of our daily life and are of vital importance for many organizations. Although the speed of this remarkable evolution has always been high, the future prospects of computer communication encompass an increasing number of new applications that are even more breathtaking than those developed in the past, including, for example, multimedia document exchange, video conferences, or cooperative work over large geographical distances. However, despite the enormous advances in physical interconnection and transmission technology, the complexity of realizing these applications is still high due to the growing demands for extensive cooperation between heterogeneous distributed systems based on increasingly powerful communication services.

Distributed computer systems use protocols to provide communication services. Improving and automating the design, implementation and testing of communication protocols in order to manage the complexity of today's and future communication services has been the scientific challenge for the work described in this book. The approach presented is an attempt to translate the view of a protocol as a set of rules into methods and tools. It reflects the author's ideas for tackling apparent problems of protocol development today, based on formal description techniques and tools. The presentation is intended to identify existing problems, characterize possible solutions, as well as positioning the proposed approach amongst others aiming at the same goals. As such, the text provides useful information to protocol developers, computer scientists interested in communication system design, as well as students at an advanced undergraduate or graduate level. For professional engineers, this book is both introductory and illustrative with respect to the principles and the application of formal methods.

The organization of the text is as follows. After giving an overview of the general architectural concepts and the development life-cycle of communication systems, existing formal description techniques are assessed to motivate the need for adopting a new approach. The technique itself, called

"Communicating Rule Systems (CRS)", is presented in detail and an operational semantics is defined. Using the operational semantics, implementation concepts and runtime mechanisms are explained that facilitate the simulated execution of CRS specifications. A possible realization of these concepts and a tool accomplishing the translation into executable form is described. Then, the general architecture of an integrated tools environment for protocol engineering is presented and the implementation of large parts of this environment is illustrated. Finally, an outlook on the future relevance of formal techniques and associated tools for the development of distributed applications is given.

Acknowledgements

This book is the published version of my doctoral dissertation, which I completed at the University of Kaiserslautern, in Germany, in 1991. The thesis was written while I was a PhD student at the IBM European Networking Center (ENC) at Heidelberg, Germany, between 1988 and 1991. During this time, I enjoyed an extremely stimulating atmosphere and was able to extend my personal knowledge considerably in a number of different ways. I am grateful to Dr. Lothar Mackert for enabling this opportunity for me, for all the inspiring conversations we had and, in particular, for his trust and friendship over the years. I am also indebted to my advisor Prof. Dr. Jürgen Nehmer. By attending his classes I learned most of what I know of distributed systems and his continuous support and guidance has always been a source of encouragement. Many thanks also go to Prof. Dr. Wolfgang Effelsberg for the help and constructive comments I received during and after his time at the ENC.

My thanks and recognition are also richly deserved by Dr. Georg Zörntlein, who went through this text several times. His accuracy, patience and technical contributions enormously improved the quality of the thesis. My ideas probably never could have been formulated in an understandable way without the generous help of Dr. Mark Mattingley-Scott, who tried hard to correct my English wording. The fact that he is already aware of my gratitude because he also had to read this acknowledgement section is a typical case of how far generosity can be exploited. Among the many other people who contributed to the successful completion of this work I would particularly like to thank my fellow PhD students at the ENC. I appreciated learning about the Dutch way of life from Rolf Velthuys during our conference visits and by sharing the same room for three years, which was a unique experience that I look forward to telling my grandchildren during cold evenings by the fire. It was also a pleasure for me to recognize that I was never observed more closely and imitated more amusingly than by Udo Bär at the celebration of my PhD. Finally, Dr. Stefan Pappe showed us all how to manage PhDs the easy way.

It was the character of the daily work at the ENC, in an international environment with students coming from a number of different countries that made the time spent there particularly rewarding. Many thanks for their contribution to my work go to B. Bär, F. Clarius, W. Eschebach, H. Janssen, T. Mangold, D. Miller, J. Monnery, S. Roorda, U. Scheere, E. Spriet and T. Stullich. Acting as their supervisor was valuable for my personal education, which was also permanently supported by Prof. Dr. Günter Müller, who headed the ENC at that time, and kept me under constant pressure to develop my abilities beyond those of an acceptable football player. I also acknowledge the help of my colleagues at the University of Kaiserslautern and the good relationship we maintained. I would particularly like to mention Ralf Reske, who organized social events very well. My thanks also go to Dr. Reinhard Gotzhein and Prof. Dr. Horst Müller for their assistance in the formal semantics part of the thesis.

My family has always been the backbone of my life. At times, they contributed more than I did to the success of this work. My parents always supported me and made my studies possible. My constant preoccupation with the problems of my thesis and the many hours I spent writing at my desk required a lot of love and patience from my wife, Rosemarie. I gratefully acknowledge her tolerance regarding my sometimes frantic behavior, as well as all the highest priority interrupts generated by our son Jens.

I dedicate this book to the memory of my cousin, Dr. Gerhard Wagner, who has been a model for me all my life and without whose encouragement I would probably have never started this work. It was God's will to call him before he could see its completion.

Ludwigshafen am Rhein, February 1992 Jürgen M. Schneider

Contents

Figures

About the author

Jürgen Mathias Schneider, son of Karl and Hildegard Schneider, was born on April 17, 1961, at Ludwigshafen/Rhein, Germany. He went to school at Ludwigshafen until completing the "Abitur" in 1980. From 1981 to 1987 he studied computer science at the University of Kaiserslautern, Germany, focussing on operating systems, communication and networking. After his diploma thesis about protocol conformance testing in 1987 he started working towards a PhD in the area of protocol engineering and the development of communication services based on formal methods and tools. In 1991 he joined IBM as a research staff member of the European Networking Center at Heidelberg, where is currently working on systems and network management.

1 Introduction

Due to the increasing availability and stability of computer networks of various sizes, more and more information processing tasks are realized as distributed applications. Ongoing advances in hardware technology facilitate the provision of decentralized computing power and enhanced communication networks. Various motivations for distributed processing exist: reduced costs, improved performance, fault tolerance, modular system design, flexibility, and extensibility [Nehm85] [Slom87].

In a distributed application, several processes running on different autonomous computer systems cooperate to accomplish a common task. To achieve synchronization and data transfer between application processes, physically distributed computer systems are interconnected by a **communication system**. The communication system comprises hardware and software equipment facilitating the provision of **communication services** to application processes. In order to realize these services, distributed components of the communication system exchange messages according to a fixed set of rules using fixed message formats. These rules and formats constitute a **communication protocol**. The activities of protocol design, validation, refinement, implementation, and testing are summarized by the term **protocol engineering**.

Efficient design and correct implementation of protocols is crucial for communication systems and for the services they have to provide. The more sophisticated these services are, the more complex are the protocol engineering tasks, but development of application programs is simplified considerably. Another aspect complicating the implementation of communication systems is heterogeneity. Connecting computer systems from several vendors, using different hardware, different architectures, and running different operating systems, poses a lot of additional problems, but also enables many new distributed applications. In order to achieve this ambitious goal, standardized communication protocols have to be used and protocol implementations have to conform to standards to guarantee interoperability.

1.1 Overview

The complexity of the services expected from a communication system have grown steadily over the last decade. Today's communication systems not only have to provide access to various physical media and cater for reliable data transfer, but also support remote operations, transaction processing, office document interchange, manufacturing messaging, and many other new functions. In general, a clear trend towards powerful application-oriented communication services can be observed.

The fact that standardized protocols are becoming more and more important puts additional burdens on implementors, who are now concerned with conformance and interoperability in heterogeneous environments. Communication hardware and software has to be developed carefully in order to achieve and maintain high quality of products. New networks employing optical fibers allow transmission speeds orders of magnitudes higher than with copper technology. These networks will trigger new distributed applications requiring new communication services and new protocols.

As a consequence, protocol engineering will become more difficult and cost-sensitive. To cope with the increasing complexity and new requirements of current and future communication networks, an improved methodology for the development of communication protocols is needed. Development and maintenance of communication hardware and software demands such a large effort that automation of development steps is desirable. Developers must be provided with a comprehensive set of powerful tools to increase the efficiency of their work and the quality of final products.

Possible automation and tool support are considered to be the main benefits that can be derived by employing formal approaches. In a formal approach, different development activities and their interrelations are defined by an underlying methodology. The fundamentals and results of development steps are documented as formal descriptions. In this way, input can be processed and output produced by tools supporting the development activities. These tools may work automatically, or under the guidance of a user.

This thesis describes the provision of tool support for protocol engineering based on a formal methodology. The initial step in this methodology is to get from an informal description to a formal specification of requirements. Further steps consist of refining and implementing the specification and finally testing the implementation. In order to facilitate these steps, suitable **formal description techniques (FDTs)** and tools have to be employed. The feasibility and efficiency of the initial step is crucial to the applicability of the whole scenario and depends on the techniques used. A rule-based formal description technique (***Communicating Rule Systems (CRS)***) is adopted in this thesis and used to illustrate basic ideas. Rules are considered to be excellent means for capturing protocol knowledge in a human-understandable way. It is shown that they are also appropriate for realizing tool support.

Formal description techniques have also been developed by standardization bodies to solve the ambiguities inherent to natural language standards. Since the beginning of protocol standardization, such ambiguities have caused many problems for implementors due to possible misinterpretations. Despite the advantages promised by the standardized FDTs, their industrial acceptance is still low. Some of the early approaches to formal description have been impressive examples of how formal techniques provide only limited help if they either complicate or oversimplify matters. Reservations of potential users are intensified by ongoing discussions of FDT experts about misconceptions that have been found in the design of the standardized FDTs.

In chapter 2, the reasons why the work conducted in the framework of this thesis was not done with one of the standardized FDTs are explained. An evaluation is presented that also includes popular non-standard FDTs[1]. The suitability of the FDTs for specifying communication services and protocols was investigated, as well as assessing properties of their underlying models. In particular, the following techniques are considered: Estelle and SDL (state machines), NPNs (Petri nets), LOTOS (process algebras), Z (predicate logic).

---

[1] The term "formal description technique (FDT)" is used in a broader scope in this thesis (detailed in chapter 2) and is not confined to standardized techniques.

Specification examples covering parts of the X.25 packet level protocol are presented to illustrate basic language features. Before discussing FDTs, informal specification aids and their shortcomings are summarized. The FDT evaluation is concluded with a comparative assessment of the techniques.

Although many of the ideas presented in other chapters of this thesis are of general nature, their realization would have been more difficult or even impossible with the techniques evaluated in chapter 2. In chapter 3, the adoption of CRS for specifying communication services and protocols is described. Modifications made to the original approach presented in [Neum88] are explained. One major modification is the replacement of CRS data type constructs by the *Abstract Syntax Notation One (ASN.1)*, as standardized in [ISO8824]. ASN.1 typing mechanisms offer considerable advantages because they are tailored to describe data structures frequently used in communication systems. Since the incorporation of ASN.1 into programming languages or specification techniques was not foreseen by the standard, several extensions to ASN.1 were necessary, resulting in an ***Extended Abstract Syntax Notation One (X-ASN.1)***. The application of both CRS and X-ASN.1 is illustrated with the X.25 packet level protocol. Formal CRS semantics is given by using *labelled transition systems* at the end of chapter 3.

How protocol engineering can be supported by tools based on FDTs is illustrated in chapter 4. E.g., various tools supporting different development activities in the formal development process can make use of executable specifications. Ideally, all tools are provided in a common software environment. However, most existing tool environments are still focused either towards protocol validation, implementation, or testing.

Chapter 5 describes protocol engineering with CRS. An operational approach is taken, based on the executability of initial and refined specifications according to the semantics defined in chapter 3. An implementation model, runtime environment, and compilation into an executable form are explained for CRS and X-ASN.1. The result of this compilation is an internal representation of the specification knowledge that is interpreted by the runtime mechanisms and provided to tools. A general architecture is proposed to integrate protocol engineering tools in a software environment that allows for use and mutual exchange of specification knowledge. The realization of such an environment based on the CRS and X-ASN.1 execution mechanisms is detailed.

The main results of the work presented in this thesis are summarized in chapter 6 and some conclusions are drawn. Possible future directions for the application of FDTs in distributed system design and realization are outlined.

The remainder of this introduction serves to clarify basic terminology and understanding of communication systems and their formal development process. In section 1.2, the layered structure of communication systems and derived notions according to the **OSI Basic Reference Model** [ISO7498] are explained. The *architectural concepts* established by this conceptual model have been subject to many discussions between implementors. In the formal approach, architectural concepts have to be considered carefully because in formal description techniques these concepts have their individual representations. A section on architectural concepts clarifies the view taken in this thesis. In section 1.3, the formal development process of communication systems is outlined. The different development activities and their interrelationship are explained. Possible tool support for each step in the process is indicated.

1.2 Communication Systems

Communication systems, as described in this thesis, are concerned with the interconnection of autonomous, physically separate heterogeneous computer systems. Whether these computer systems represent a loosely coupled computer network or form a *distributed system* is not relevant to the investigation of providing complex communication support. Thus, the general discussion about what constitutes a distributed system (see for example [Ensl78] [Slom87] [Tane88] etc.) is outside the scope of this thesis.

A general model for communication systems which has gained much recognition and has become widely accepted was published by the International Organization for Standardization (ISO) in 1984, called the *Open*

Systems Interconnection (OSI) **Basic Reference Model** [ISO7498]. Systems that use the internationally standardized protocols of OSI in their communication with other systems are called **open systems** because they are 'open' to connect to any other open system in the world. The structuring principles and basic architectural concepts of communication systems according to the OSI Basic Reference Model are introduced in section 1.2.1. The interpretation of architectural concepts is further detailed in section 1.2.2.

1.2.1 Structuring Principles

A typical configuration of open systems is depicted in Figure 1. A part of each open system is responsible for data processing, whereas other parts offer data communication services. The latter are connected to physical media in the form of various communication networks, ranging in size from

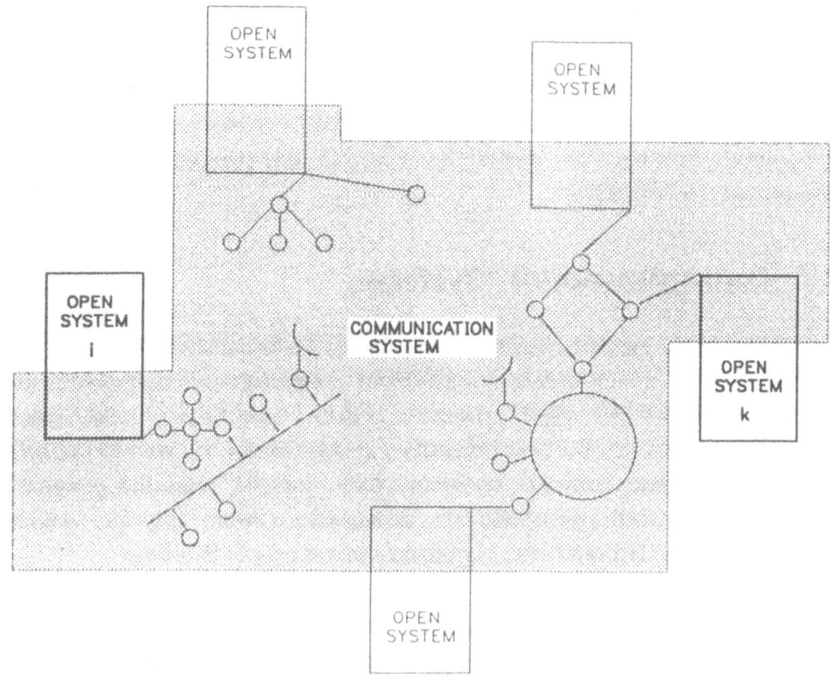

Figure 1. Communication system interconnecting open systems

local area networks (LANs) to metropolitan and wide area networks (MANs, WANs) (see for example [Tane88]).

The separation between data processing and data communication tasks in an open system is further detailed in Figure 2. A number of **application processes** realize information processing for the distributed application. Communication services are provided to application processes by **communication subsystems** in each open system. The collection of all communication subsystems yields the complete communication system[2].

In order to reduce complexity, communication systems are modelled as *layered systems*. The whole communication system is decomposed into a hierarchy of layers of manageable complexity. Each layer consists of a number of distributed entities that make use of the services of the next lower layer to execute a communication protocol and to provide enhanced services to the layer above. The first step of this decomposition is shown in Figure 3 (compare [Viss86b]), yielding a number of **application entities** in

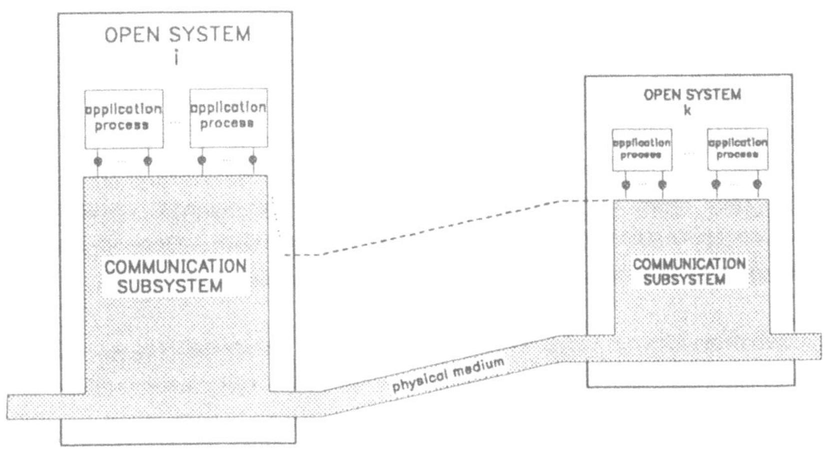

Figure 2. Communication system as a collection of subsystems

[2] In Figure 2, and in following figures, dotted lines are used to indicate logical rather than physical interconnection.

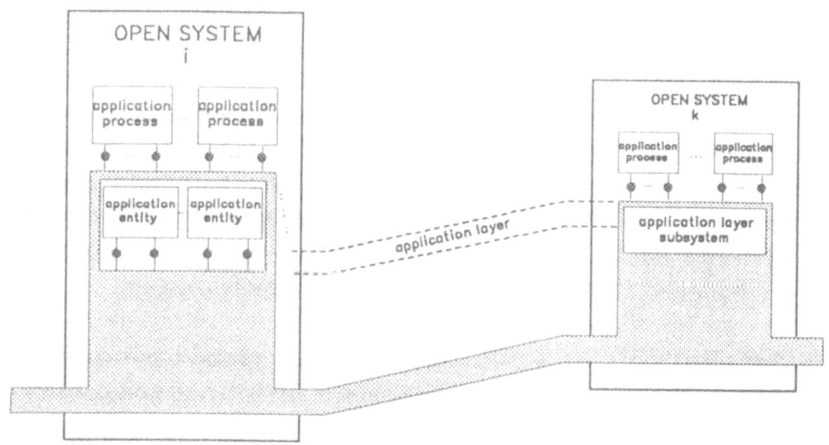

Figure 3. Decomposition of communication services

each open system and underlying lower layers. The collection of all application entities in an open system constitutes the **application layer subsystem**. Application layer subsystems existing in each open system together form the **application layer**.

The decomposition process is continued layer by layer. For reasons of simplicity, the number of layers should not get too large. The OSI Basic Reference Model divides a communication system into seven layers and establishes a framework for additional standards describing the individual layers (see Figure 4). Readers not familiar with the purpose of each layer are referred to [ISO7498], or may study the individual layer standards for complete information. A very good overview is provided in [Effe86].

Further modelling is required for the interaction of entities situated in adjacent layers and for the communication between distributed entities of the same layer. To explain these concepts, an arbitrary single layer is regarded in isolation and referred to as the **(N)-layer**. The number N indicates the position of the layer in the whole architecture. The purpose of an (N)-layer is to provide the **(N)-service** to the next higher layer (**(N+1)-layer**), as depicted in Figure 5. To achieve this task, several **(N)-entities** exchange

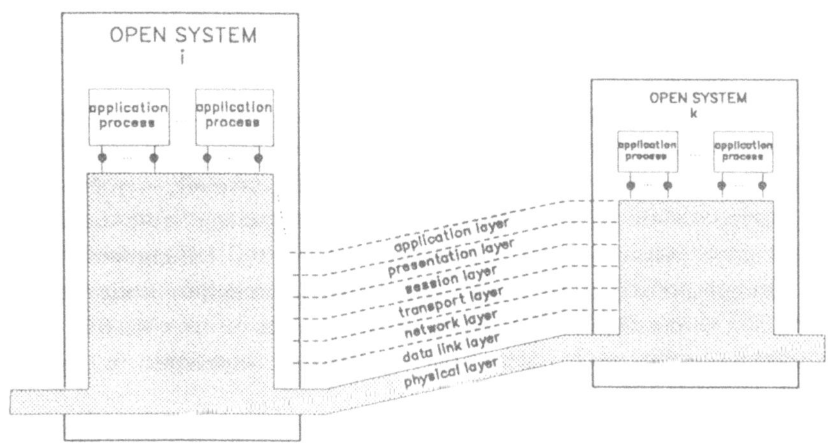

Figure 4. Layered communication system as defined by the OSI Basic Reference Model

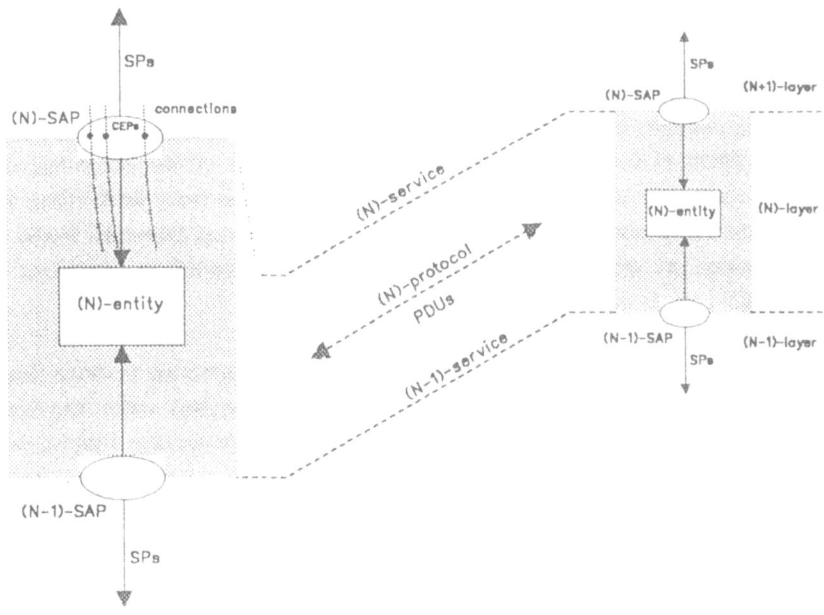

Figure 5. An (N)-layer in a layered communication system

information according to rules defined by the **(N)-protocol**, using the service of the next lower layer (**(N-1)-service, (N-1)-layer**).

An (N)-service is provided by the (N)-layer at one or more **(N)-service access points** ((N)-SAPs) in form of a set of **service primitives** (SPs). An (N)-SAP is associated with an **(N)-address** that uniquely identifies it; SPs exchanged at SAPs may convey parameters, but nothing is stated about the subtleties of their realization in real systems. Several **(N)-connections** may run through one (N)-SAP and are identified by **connection endpoint identifiers** (CEPs) that have to be unique in the scope of one (N)-SAP. An (N)-connection with more than two connection endpoints is called a **multi-endpoint connection**.

Entities situated in the same layer, but located in different open systems, are called **peer entities**. Peer entities communicate by exchanging **protocol data units** (PDUs) of well-defined format, according to the (N)-protocol. PDUs are not directly transmitted (except for the physical layer), but passed to the next lower layer as an SP parameter. Entities situated in adjacent layers, but located in the same open system, interact in terms of **interface data units (IDUs)**. In an IDU, upper layer PDUs are included in a **service data unit (SDU)**, together with some **interface control information (ICI)**. The lower layer entity interprets the ICI and adds some **protocol control information (PCI)** to the SDU to form the lower layer PDU. At the receiving side, the PCI header is stripped off and interpreted by the peer entity. The SDU information is passed to the higher layer. The mapping between these different kinds of data units in communication systems is visualized in Figure 6.

There are many problems that may have to be overcome in each layer. *Flow control* mechanisms are usually required for entities within the same layer and between entities in adjacent layers. *Sequencing* mechanisms have to be present if the underlying layer does not guarantee to deliver data in the same order as submitted. In case of corruption, duplication, or loss of data, *error detection* must be assured. Multiple connections in one layer may be mapped onto one connection in a lower layer (*multiplexing/demultiplexing*), or several connections in a lower layer may be used to support one connection in a higher layer

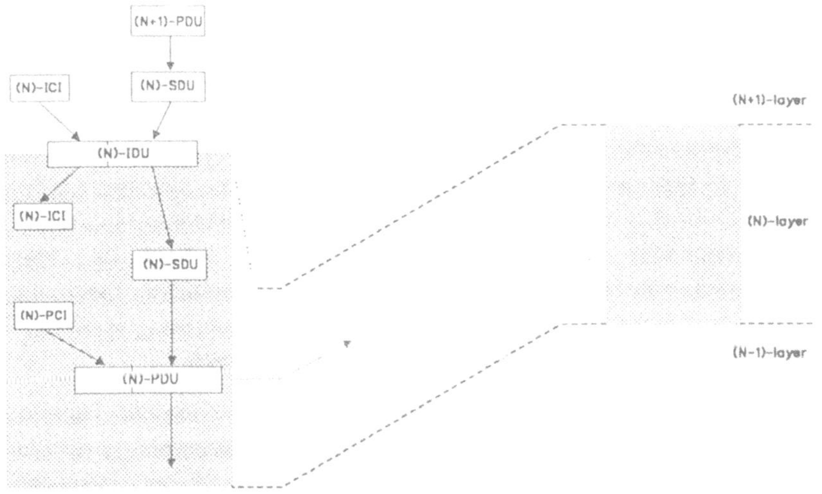

Figure 6. Mapping between data units in adjacent layers

(*splitting/recombining*). One (N)-SDU may be mapped onto several (N-1)-PDUs (*segmenting/reassembling*), or multiple (N)-SDUs fit into one (N-1)-PDU (*blocking/deblocking*). Finally, multiple (N)-PDUs may be mapped onto one (N-1)-SDU (*concatenation/separation*).

The introduction of architectural concepts to model OSI, as summarized in this section, was one of the major contributions of the OSI Basic Reference Model. Although these notions are widely used now, [ISO7498] fails to give precise meaning to all of them [Viss86b] [Svob89] [Gotz89]. The intention was to stay on a high level of abstraction and leave as much freedom as possible for implementors, but as an effect, vague and ambiguous interpretation of these elementary notions has led to differing realizations. In the next section, the interpretation of architectural concepts used in this thesis is described. Providing precise definitions is, however, difficult without a formal framework. The discussion should serve to reveal basic properties and to establish a basis for assessing representations of architectural concepts in FDTs.

1.2.2 Architectural Concepts

The concepts and notions summarized in the previous section are basic architectural concepts[3] of communication systems. It is emphasized in [ISO7498], that the OSI Basic Reference Model should not be interpreted as an implementation specification, but as a conceptual framework. However, the level of abstraction at which the architectural concepts are introduced, and their use in the natural language text, allows differing interpretations. These differences run the risk of incompatible choices by different implementors [Svob89]. In order to avoid misinterpretations, clear understanding of architectural concepts is of utmost importance [Gotz90a]. The basic architectural concepts can be described as follows:

- A communication system consists of a hierarchy of **layers**. Each layer can be regarded as an abstract machine that implements a set of functions to provide a service to its user. Service users are only interested in the functionality provided by a layer, not in the way it is implemented. Several layers may run in parallel and may be realized as one or several system processes, collections of modules and procedures, or may be built in hardware.
- The functionality of a layer is described by its **service** definition. A service is defined by

  1. a set of service primitives (SPs) exchanged at service access points (SAPs);
  2. restrictions on the local ordering of SPs at one SAP;
  3. restrictions on the global ordering of SPs between two or more SAPs;
  4. dependencies of SP parameters.

  Some model of the internal behavior of a layer may be used to describe the service. However, only external behavior in terms of SPs is relevant for the service definition.

[3] The following general definition of *architectural concepts* is given in [Brin88a]: "The class of informal and widely used objects and operations that are subject to specification".

- The **service primitives (SPs)** are the input/output events of a layer. To exchange an SP, an interaction between a service provider and a service user is required. Each SP may convey a number of parameters that are passed from a service user to a service provider and vice versa. SPs may be realized as procedure calls, system calls, interrupts etc.

- **Service access points (SAPs)** symbolize the location where SPs are exchanged between a service provider and a service user. The semantics of communication at SAPs is not fixed. In principle, synchronous or asynchronous communication mode is possible. Properties of SAPs may comprise reliability, capacity, order preserving, context-dependency etc. (compare [Gotz89]).

- The active elements in a layer are called **entities**. Entities are responsible for realizing layer functions. They communicate with peer entities in other systems by using a communication protocol. To achieve this communication, each entity is connected to one or more SAPs of the lower layer and uses the lower layer service. Enhanced services are provided at SAPs of the next higher layer. The number of entities in a layer is not fixed; entities may also execute in parallel. Protocol entities may be realized as single processes, procedures, or as part of those.

- A **protocol** is a set of rules and data formats governing the communication between peer entities. These rules and formats constitute a distributed algorithm facilitating the provision of the layer service. A protocol is derived from the service by decomposition into a number of protocol entities and an underlying service. A protocol is specified by

    1. the definition of the service to be provided;

    2. the definition of the lower layer service;

    3. the definition of PDU formats both in an abstract (decoded) form, as well as a bit-level representation (encoded form);

    4. rules describing the mapping of SPs and parameters to PDUs and PDU fields and vice versa, the admissible sequences of events within the layer, and the associated internal protocol actions;

    5. rules describing protocol actions upon internal conditions (timeouts, buffer limitations, hardware errors etc.).

Protocol actions are usually implemented by programming language statements or hardware instructions. Protocol data formats are implemented by programming language type definitions, or generated and recognized by dedicated hardware circuits.

- Associations between one or more higher layer entities established by the layer underneath are called **connections**. Normally, protocols and services allow for several connections running through a single SAP. Connections are means to decide which entity is to be addressed when sending and receiving data. Connections may allow for point-to-point, multicast, or broadcast communication.

As may have become apparent from this discussion, the definition and understanding of architectural concepts for communication systems is a difficult subject due to the high degree of abstraction required in the framework of open systems. The application and use of FDTs has stimulated and influenced international agreement on semantics of architectural concepts. The representation and meaning of architectural concepts in different FDTs is an important issue.

1.3 Protocol Engineering

Software development in general was considerably improved both in terms of quality and productivity by more rigorous approaches to software engineering. One of the results was the identification of a so-called "software life-cycle" in which program development was split into different phases, known as requirements analysis, design, implementation, and testing (see for example [Lisk86]).

The increased complexity and new requirements of open systems revealed weaknesses in the informal approach to software engineering for communication protocols. Most of the problems resulted from ambiguities in the initial requirements description, incompleteness and incorrectness of the design specification, structural deficiencies in the implementation, and limitations of ad-hoc approaches in the final testing phase.

A formal development methodology not only defines several consecutive development steps, but relates the input and output of each step by means of mathematical notions (for example, functions and relations). In this way, the meaning of development steps is fixed. During the development services and protocols have to be described at varying levels of detail in order to suit different needs. The role of FDTs in the development process is crucial since they are the key to providing powerful tool support for each single step. Tools facilitate rigorous analysis of a design, automated validation, code generation, and test derivation.

Formal methods have proved to be fruitful for protocol engineering, although their application has turned out to be difficult [Boch80a] [Rudi88]. One reason was, and still is, a lack of education and training in the use of formal techniques and in the conduction of a formalized methodology, as outlined in section 1.3.1. Another reason is that it is only after spending time and effort in fundamental research that common agreement on such a formal methodology is beginning to emerge. This leads to an improved understanding and better design of tools, summarized in section 1.3.2, that are intended to support the different development activities on the basis of formal descriptions.

1.3.1 Life-cycle

The general formal development life-cycle is presented in Figure 7 (see also [Boch87a] [Brin88a] [Drob89] [Turn90] [Sidh90a]). The initial task in this life-cycle is the definition of system requirements. Clear understanding of desired functionality has to be reached. Then an architecture may be designed that satisfies the requirements. This is either done in a completely informal way using natural language and pictures, or semi-formally by applying graphic- and table-based techniques.

The next goal is the ***formalization*** of the system requirements, resulting in a high-level formal specification describing the functional behavior of the system. In communication systems, this specification is referred to as the *service definition*. It precisely defines all possible sequences of external interactions with service users.

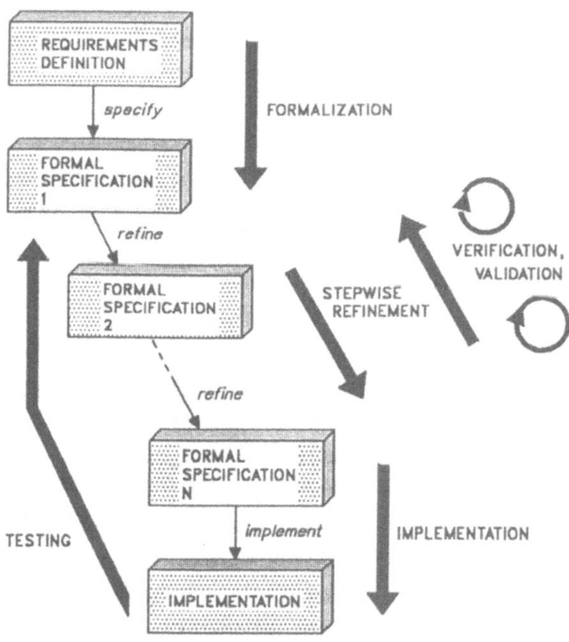

Figure 7. Formal development process

The availability of a formal specification facilitates rigorous analysis of the specified system. The activity of ***validation*** aims at *checking* specific properties as given in the requirements definition and investigating general system properties, such as consistency, completeness, and absence of deadlocks. Another approach is ***verification***. Here, the goal is to *prove* that the specification is free of errors and guarantees all the desired properties in all situations. The term verification is used in a sense according to [Boch80a] [Lisk86], involving logical reasoning based on the specification text.

Usually, the initial high-level specification cannot be implemented directly. It has to be refined structurally (for example, into a number of components) and/or functionally (resolving options, adding internal details), maintaining some form of equivalence to the top-level specification [Brin90]. In general, this activity may be repeated several times and is called ***stepwise refine-***

ment. Thus, a communication *service definition* is refined into a *protocol specification* and further into an implementation-oriented *system specification.*

The final specification reached has to be realized on a physical system. This activity is called ***implementation*** and its outcome is the goal of the whole process. If a sufficiently high level of granularity has been reached, implementation structures may directly reflect the structure of the final system specification.

The remaining activity is ***testing***. Testing is important because of the technical or practical infeasibility of formal verification during all steps of refinement and implementation. It can be applied to the validation of intermediate specifications against their predecessors, or checking implementations against specifications. Testing an implementation against the initial specification means performing "black-box" testing, referred to as *conformance testing* in ISO standards [ISO9646]. Other forms of testing, for example, "grey-" or "white-box" testing, require knowledge about the internal structure or realization of the system components, which is available in some of the intermediate system specifications.

Finally, all or part of the development process has to be reassessed for the purpose of maintenance, modification, performance improvement, or addition of system requirements.

1.3.2 Tool Support

Tools usage during requirements analysis is largely restricted to general purpose tools, such as word processing systems or graphic workstations. They help to produce natural language documentation containing figures, diagrams, and tables. Semi-formal techniques using a fixed set of graphical symbols with an associated meaning may also be appropriate. By employing formal techniques, system descriptions are unambiguous. After the ***formalization*** step, architectural concepts like entities, interaction points, connections etc., used informally during requirements analysis, have concrete representations in the formal language chosen [Gotz89] [Brin88a]. Language constructs used for the description have precise syntax and semantics and can be processed by tools. Tools used during the formalization process are structure-based or graphical editors, parsers, and static analyzers, which check language syntax and static semantics.

Specification ***validation*** tools are dynamic analyzers that can be classified into *exhaustive analyzers* and *simulators*, according to [Boch87a]. They use mechanisms for *interpretation* or *symbolic execution* of specifications, based on formal semantics of the specification language (see for example [Logr88] [Saqu89]). A well-known validation method for state-oriented techniques is *reachability analysis*, which works very well provided the state space is limited. However, the problem of "state space explosion" affects its practical usefulness for many real protocols. Simulative validation is often carried out interactively by the protocol designer checking the most critical paths in the specification. More thorough simulations may be set up and guided by heuristics or probabilities [West87].

The activity of ***verification*** may be supported by theorem provers or proof assisting tools [Aujl89] [Boot90], but even then the activity remains very time-consuming and often infeasible for the specification as a whole. Sometimes, it is applied for the most critical sections, or employs a simpler model that may be sufficient for verifying basic properties.

Stepwise refinement can be guided by a set of transformation rules and may be automated by tools implementing these rules. Automation of this activity is approached by inventing various *specification styles* together with theories about semantic-preserving transformations between them. At the time being, considerable complexity is faced when realizing such tools (see for example [Eijk90a]) and the identification of styles is often a matter of debate. However, validation of refined specifications would become obsolete if refinement was done automatically and preserved some form of equivalence. A weaker form of tool support for stepwise refinement is given by tools assisting an interactive designer. Such tools can make proposals, based on general design rules, that have to be assessed and modified by the interactive user.

The activity of ***implementation*** can be supported by code generators if the final specification has reached an adequately low level of abstraction [Boch87b] [Vuon88] [Flei89] [Eijk90b]. The amount of code that can be generated automatically varies from simple code frames to complete control structures. Recent activities show that the implementation of protocol data structures and encoding/decoding functions may also be automated (see for example [Gora88a] [Sche89] [Neuf90]). Parts of the overall code depending on operating system, communication subsystem, memory management, or application-specific interfaces have to be added by hand. Tools for the implementation step usually rely on language-specific translators and compilers. Support packages for interprocess communication, buffer allocation, interrupt handling etc. can be used.

As has been recognized quite early, tool support is of particular importance in the area of ***testing***. Meaningful tests may only be performed and evaluated against some formal specification. Several theories for the derivation of tests have been developed and implemented in corresponding tools based on the models of FSMs, EFSMs, Petri Nets, and process algebras [Sidh89] [Ural87] [Baum90] [Alde90]. These tools generate test sequences and test data, criteria for test selection and verdict assignment, and also allow test execution (see for example [Chan90] [Stol90]). However, a lot of problems concerning, for example, test data selection, test coverage, and test purpose generation still have to be solved [Dwye90].

2 Protocol Specification with Formal Description Techniques

At present, various means have been proposed for specifying services and protocols for the different layers in a communication system. The specifications form the basis for design validation and serve as references for implementors. For validating the correctness of implementations, tests can be derived from specifications. According to [Lisk86], a specification can be considered an abstraction from a set of implementations. The primary purpose of the abstraction is to focus on the *functional requirements* of a system that can be met in various ways by possible implementations [Viss86b]. To achieve this purpose, specifications need to be general enough so that they do not exclude admissible implementations unnecessarily, but also have to be restrictive enough to rule out unacceptable ones. Desirable properties of specifications include preciseness, consistency, completeness, and structuredness (compare [Brin88a]).

For a long time, the architectural concepts of communication systems were not described as rigorously as today. Many specifications focussed on the protocol, without taking service aspects or interactions between entities in a stack of protocol layers into account. With the definition of the 7-layer OSI Basic Reference Model a conceptual framework has been established that provides the terminology and background for specifying the services and protocols of individual layers in a communication system. The architecture and terminology of the Basic Reference Model are used in the ISO and CCITT specifications of the OSI layers. These specifications are expressed by means of natural language text, augmented with pictures, tables, and diagrams. Such informal specification elements are illustrated in section 2.1 by briefly summarizing the X.25 packet level protocol (PLP), which is then used as an example throughout this thesis.

Formal modelling of services and protocols was mainly motivated by the fact that design errors in natural language specifications or misinterpretations due to ambiguities could only be detected late in the implementation process. In order to correct these errors, large parts of existing programs had to be rewritten, which was very costly. A suitable formal model allows validation of the design and provides a precise reference for implementation. The most popular formal models applied to protocols include state machines, Petri nets, process algebras, and models based on logic. Complete **formal description techniques (FDTs)** have been developed based on these models to provide intellectual tools for specification. In section 2.2, FDT candidates being based on different models are investigated to assess their suitability for specifying communication services and protocols. A final comparison of the evaluated techniques is provided in section 2.3.

2.1 Informal Specification

Specifications expressed by means of natural language, pictures, diagrams, and tables are called *informal* specifications because these specification mechanisms have no fixed syntax and no formally defined meaning. Natural language descriptions of communication protocols explain the position and objectives of the protocol (probably by relating it to the OSI Basic Reference Model), before focussing on the details of how the objectives are met. The PDU formats and the meaning of individual fields are described. The appropriate situations for sending or receiving PDUs are explained and the allowed values of PDU fields are indicated. The relationship between the protocol and the service is established by natural language text and illustrated by figures.

Some typical elements of an informal protocol specification are recalled in this section using the X.25 packet level protocol (PLP) as an example. The CCITT (Comité Consultatif International de Télégraphique et Téléphonique) recommendation X.25 was first defined in 1976 and has been revised by CCITT several times until today. The X.25 PLP has been adopted by ISO in [ISO8208] and is today a widely known connection-oriented network layer protocol used by many national carriers and private institutions. It serves for managing communication between two user *DTEs* (*Data Terminal Equipment*) from the CCITT terminology. Each DTE is connected to a *DCE (Data Circuit-terminating Equipment)* providing access to a packet-switched

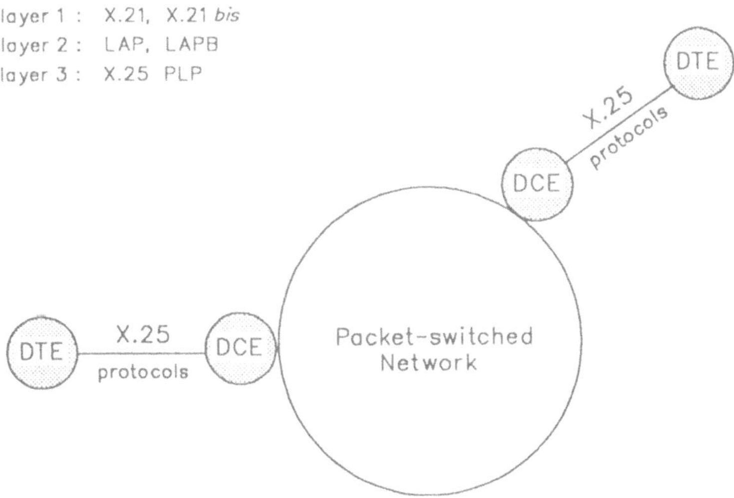

Figure 8. X.25 protocols

network[4] (see Figure 8). Together with the local and the remote DCE, the network can be regarded as the underlying service provider passing protocol messages between DTEs. These messages are called **packets** in the X.25 PLP. Two forms of connections are supported by the protocol, namely **switched virtual calls (SVCs)** and **permanent virtual circuits (PVCs)**. SVCs are set up before data transfer and are cleared afterwards. PVCs are similar to a leased line and require neither setup nor clearing. A so-called **logical channel** is associated with each SVC and PVC by the protocol. The use of the X.25 PLP to provide the OSI connection-oriented network service (OSI-CONS) [ISO8348] has been standardized in [ISO8878].

The relationship between the OSI-CONS and the X.25 PLP is illustrated in Figure 9 [ISO8878]. Network service users (transport entities) interact with network entities in the underlying layer (network service provider) in terms

---

[4] The DTE-DTE case is not used in this thesis.

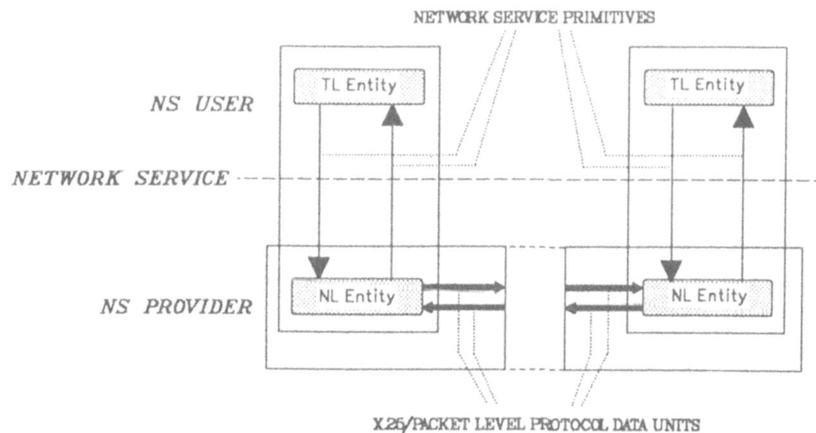

Figure 9. Use of X.25 to provide the OSI connection-oriented network service

of network SPs. The network entities exchange packets (X.25 PDUs) according to the X.25 PLP in order to provide the service. However, there is no precise semantics behind the pictorial elements, that, for example, defines the possible parallelism between the different entities and describes their interaction and communication. This may lead to misinterpretations.

As an example consider the situation where the network service user issued an N-CONNECT-request primitive and an INCOMING CALL packet arrives. Depending on the actual sequence of events the network entity may have passed the CALL REQUEST packet to the DCE and a "call collision" occurs. In [ISO8878], paragraph 6.2.1, it is stated: "When an NL entity receives an INCOMING CALL ... packet, it signals an N-CONNECT-indication ... primitive ... to the NS user.". However, in the "call collision" case this violates the network service. The text accompanying Figure 9 in [ISO8878] is incomplete with respect to this situation.

Figure 10 summarizes the network SPs. In standards, such tables are also used to list SP parameters and to indicate whether parameters are mandatory or optional (see [ISO8348]). The types of SP parameters are described informally by enumerating possible values or characterizing possible ranges. The X.25 PLP packet types are given in Figure 11. In [ISO8208], the

connection establishment	N-CONNECT-request N-CONNECT-indication N-CONNECT-response N-CONNECT-confirm
data transfer	N-DATA-request N-DATA-indication N-DATA-ACKNOWLEDGE-request N-DATA-ACKNOWLEDGE-indication N-EXPEDITED-DATA-request N-EXPEDITED-DATA-indication N-RESET-request N-RESET-indication N-RESET-response N-RESET-confirm
connection release	N-DISCONNECT-request N-DISCONNECT-indication

Figure 10. Network service primitives

connection establishment (call setup)	CALL REQUEST INCOMING CALL CALL ACCEPTED CALL CONNECTED
data transfer	DATA INTERRUPT INTERRUPT CONFIRMATION RECEIVE READY RECEIVE NOT READY REJECT RESET REQUEST RESET CONFIRMATION
connection release (call clearing)	CLEAR REQUEST CLEAR INDICATION CLEAR CONFIRMATION
restart	RESTART REQUEST RESTART CONFIRMATION DIAGNOSTIC REGISTRATION REQUEST REGISTRATION CONFIRMATION

Figure 11. X.25 packets

structure of each packet is described by pictures similar to the ones shown in Figure 12. Additional information about the types of packet fields and their individual encoding is given in the text.

Figure 12 shows the format of the CALL REQUEST packet (left-hand side of the figure). The parts drawn in bold lines represent the packet header contained in all X.25 packets in the first three bytes. The first four bits constitute the so-called "general format identifier", which is constant for all packets except DATA packets. The number of the logical channel the packet belongs to is contained in the second half of the first byte (group number), and in the second byte of the packet header (channel number). The 12 bits provided allow for 4096 logical channels. However, use of channel number 0 is reserved. The third byte contains the packet type identifier, which is an individual bit pattern for each packet. The "general format identifier" in DATA packets contains important information for the DTEs executing the X.25 PLP (see right-hand side of Figure 12). The "Q-bit" can be used to distinguish between qualified and non-qualified data, the "D-bit" can be used for requesting receipt confirmation of data packets. The numbering scheme

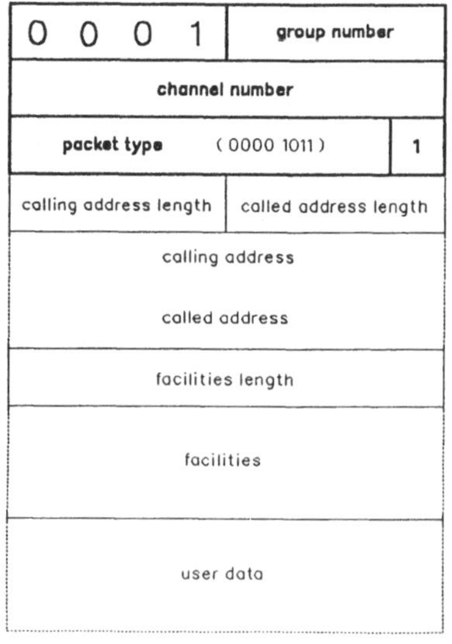

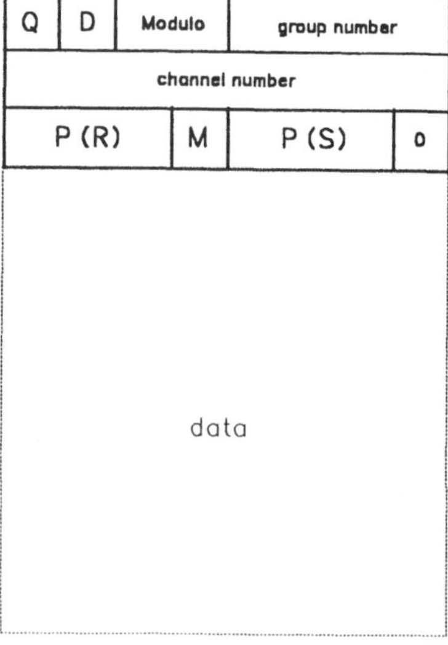

Figure 12. X.25 PLP CALL REQUEST and DATA packet formats

used for packet sequence numbers (modulo 8 or modulo 128) is encoded in the following two bits. The packet type identifier of DATA packets is encoded in 1 bit which is set to zero (for all other packets, the last bit of the packet type identifier is set to one). The remaining seven bits contain the number of the next data packet expected ("P(R)"), the "M-bit" indicating whether more data follows, and the send sequence number of the packet ("P(S)"). Sequence numbers allow transmission errors to be detected, for example, loss or duplication of packets. A sliding window is maintained separately for both directions of data transfer to regulate the flow of data between a DTE and a DCE.

An example of ambiguity in the informal X.25 PLP specification is the following: the dotted lines in Figure 12 indicate that PDU fields have variable length. This includes length "0" and implies that both addresses may be absent in the CALL REQUEST packet. In [ISO8208], paragraph 5.2.1, it is stated: "The CALL REQUEST packet may include the called-DTE address and the calling-DTE address.". In fact, it cannot be precisely deduced from neither the text, nor from the picture, whether no addresses, only one address, or both addresses are all legal alternatives.

During an execution of the X.25 PLP, DTEs are in different states. In each state, several packets may be sent or received. In order to define the possible incoming and outgoing packets for all states of the protocol, *state diagrams* are used. In Figure 13, the state diagram for the X.25 PLP call setup phase is shown. From a service user's point of view, internal protocol states are not of interest, but possible sequences of SPs are. An explicit representation of sequence information is provided by *time sequence diagrams*. In Figure 14, the time sequence diagram for call setup in the X.25 PLP is shown. Vertical bars are used to illustrate the interfaces that data units have to pass through and arrows indicate data flow. Time proceeds from the top to the bottom of the diagram. The connection establishment is initiated by a network service user with an N-CONNECT-request primitive. A CALL REQUEST packet is sent by the network entity and delivered as an INCOMING CALL packet to the receiving side. An N-CONNECT-indication primitive is passed to the remote network service user, who responds with an N-CONNECT-response primitive. A CALL ACCEPTED packet is sent and delivered as a CALL CONNECTED packet. An N-CONNECT-confirm primitive is

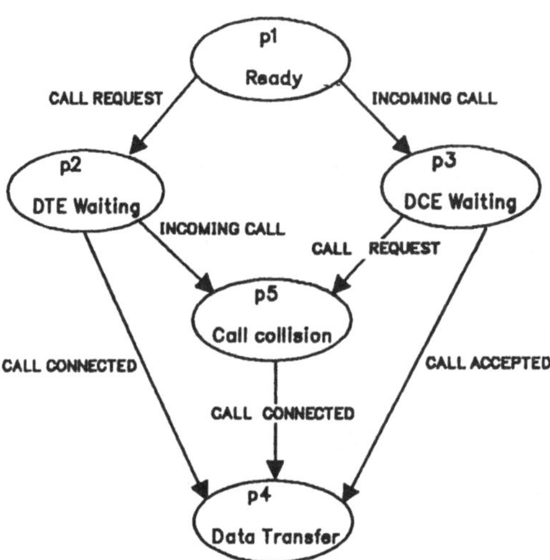

Figure 13. X.25 PLP call setup state diagram

passed to the network service user, signalling that the requested connection has been established.

Figure 14 describes partially the mapping between network SPs and X.25 packets according to [ISO8878]. The CCITT X.25 PLP had been defined

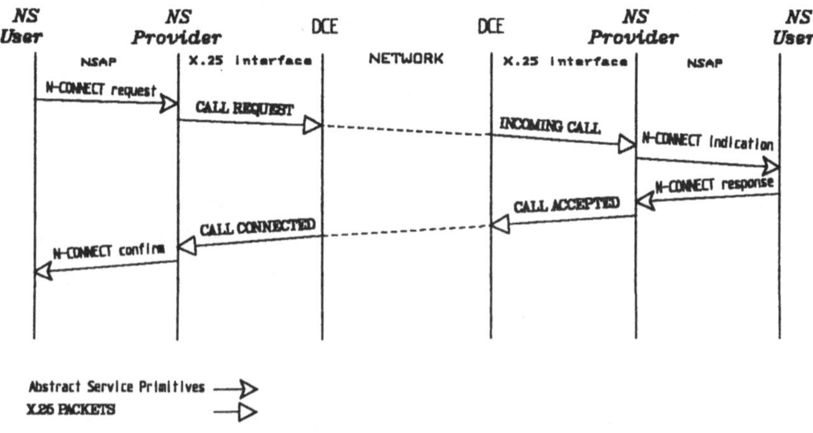

Figure 14. Time sequence diagram for call setup in the X.25 PLP

before the ISO network service and, consequently, before the mapping between the service and the protocol. This does not correspond to the development cycle as presented in chapter 1, but has historical reasons. The mapping of SP parameters to packet fields and vice versa is given in several individual paragraphs and clauses of English text in [ISO8878]. The meaning and intended use of packet fields has been described using many sentences spread over different sections of the document [ISO8208]. In general, such natural language descriptions are easy to read, but as already illustrated, readability does not necessarily induce intelligibility. In fact, ambiguities cannot be avoided in natural language text and important information may be missing. Besides these shortcomings, the fact that informal specifications cannot be processed by tools supporting validation, implementation, and testing of services and protocols is a major drawback. Formal models and FDTs have been invented to overcome this problem.

2.2 Formal Description Techniques (FDTs)

A description of an object is a symbolic representation of the object in some description language. Formal descriptions are descriptions by means of a language that possesses the following two properties:

1. a **formal syntax**, which is a set of rules prescribing all the symbols that can be used and the expressions that can be constructed in the language;
2. a **formal semantics**, which are rules for the unambiguous interpretation of language expressions on the basis of a mathematically defined semantical model.

Such languages are called **formal description techniques (FDTs)** (compare [Viss86b]) and are used to produce formal specifications.

The requirements for specifications explained at the beginning of this chapter demand corresponding properties for FDTs. Several existing FDTs have been investigated in the framework of this thesis. In order to judge the suitability of these FDTs for specifying communication services and protocols, the following general criteria have been used (compare [Brin88a] [Bolo87] [Brin87] [Viss86b]):

- ***formal definition***, which refers to the existence of a formal syntax and a formal semantics. The underlying formal model of an FDT should facilitate the development of analytical theories for validation, implementation, and testing, as well as simplifying tool development;

- ***expressiveness***, which demands language constructs that allow convenient description of a wide range of properties or functionality. The "modelling power" of a technique is important, but also its practical applicability for specifying complex systems;

- ***abstraction***, which means to what extent implementation-independent specifications are supported by the underlying model and the constructs of the language. Since only functional requirements should be defined by specifications, the technique should be free of details that are only relevant to implementations;

- ***compositionality***, which requires language features for composing individual parts of a specification in various ways to form a whole system. If systems are complex, it should be possible to decompose them into smaller units that can be described separately. However, it must still be possible to deduce the overall behavior of the system from the behavior of the components;

- ***structure***, which means that specifiers should be provided with mechanisms to organize formal descriptions in an adequate way. It should be possible to aggregate related aspects, separate unrelated ones, distinguish phases of different behavior, or support readability by defining a suitable structure;

- ***intelligibility***, which demands that language constructs and underlying models are simple enough to be understood after a relatively short training period. To be applicable in practice, FDTs must be understandable for all members of a development team and not only for a few formal description experts.

Another aspect which is particularly important for the ideas presented in this thesis is

- ***constructivity***, which is the ability to interpret formal descriptions algorithmically[6]. Constructive formal descriptions can be executed in terms of the underlying semantical model (operational specification). In this way, the specification directly leads to an executable system satisfying the functional requirements.

Besides these general criteria, the intended application area must be kept in mind when evaluating the suitability of FDTs. For example, powerful constructs for the parallel composition of specification parts do not contribute to the suitability of a technique for describing sequential systems, whereas for specifying concurrent activities they are crucial. In general, the formal semantics of an FDT should be unambiguously related to the architectural concepts of the intended application area [Viss86b] [Turn87] [Gotz90b]. In [Turn87], the link between FDT concepts and architectural concepts of the application area is referred to as the *architectural semantics* of an FDT.

The evaluation of several existing FDTs with respect to the general criteria and with respect to the application area of communication systems is presented in the following sections. The selected FDTs use different underlying models and have been applied frequently to the specification of communication services and protocols. To illustrate the basic characteristics of the selected FDTs, the underlying models are explained first. Setting up a virtual call in the X.25 PLP is used as a common example and partially described in all techniques.

2.2.1 Estelle and SDL

The formal description technique Estelle was developed within ISO and became an international standard in 1988 [ISO9074]. It is intended to be used for specifying distributed concurrent systems and, in particular, OSI services and protocols. An elaborate tutorial is provided in [Budk87]. The development of SDL (*Specification and Description Language*) began within CCITT in 1968. In order to provide a specification tool for telecommunications equipment to be used in many national and private institutions, a

---

[6] In [Brin88a], an FDT is called constructive if there exists an interpretation algorithm for the symbolic execution of specifications.

pragmatic approach was taken. As an effect, different versions of the technique had been in use many years before SDL was standardized in 1987 [CCZ100]. Both Estelle and SDL are based on the model of *extended finite state machines (EFSMs).*

Extended Finite State Machines

Several notions of the concept of state machines exist (see for example [Hopc79]). The ones most suitable for modelling protocols are state machines with inputs and outputs. A **finite state machine (FSM)** with inputs and outputs is defined as

$M=(Q, I, O, \delta, \lambda, q)$, where

Q	is a finite set of states,
I	is a finite set of inputs (input alphabet),
O	is a finite set of outputs (output alphabet),
$\delta: Q \times I \rightarrow Q$	is the transition function.
$\lambda: Q \times I \rightarrow O$	is the output function,
$q \in Q$	is the initial state.

In this definition, outputs are associated with transitions and depend on the input and the current state (*Mealy* machine). Alternatively, the output function λ may also be defined as $\lambda: Q \rightarrow O$, associating outputs with states (outputs depend on states only, *Moore* machine). Mealy and Moore machines are deterministic: subsequent states are uniquely determined by the current state (and the input). If the transition function δ is defined as a relation $\delta: Q \times I \rightarrow P(Q)$ ($P(Q)$ denoting the powerset of Q), the state machine is called nondeterministic. FSMs are graphically represented as state diagrams (see for example Figure 13) or as state tables.

When applied to protocol specification, input and output symbols of an FSM represent SPs or PDUs and states model internal states of a protocol entity. A state remembers the input/output history of an FSM in so far as it is needed to determine subsequent behavior. In this way, possible input/output sequences of an FSM are described. However, the FSM specification does not cover all protocol aspects, but focuses on the sequencing of events and protocol actions (protocol behavior). Additional information is usually given by natural language text or by using extended models.

The model of an **extended finite state machine (EFSM)** is derived from initial ideas presented in [Boch77]. In an EFSM, states are augmented with *context variables* that do not "contribute" to the FSM. To differentiate state objects, the terms *major state variables* and *minor state variables* are also often used. The inputs and outputs of an EFSM are associated with *parameters*. The transitions of an EFSM depend on the major state and the input, but also on *predicates* over minor state variables and input parameters. Additionally, transition *actions* may alter values of minor state variables and assign values to output parameters.

For the purpose of communication, several mechanisms have been proposed for interconnecting state machines. Direct coupling of transitions in adjacent state machines is one possibility. At the other extreme, there is the concept of a *channel* existing between a pair of state machines. Channels may or may not have buffering capacity. If buffering of interactions takes place, queueing models are often employed (for example, FIFO queues).

Estelle

An Estelle specification describes a hierarchically ordered set of EFSMs, called **module instances** (or **tasks**), running in parallel and communicating via message passing. Minor state variables, predicates, and actions of EFSMs are represented as "PASCAL" variables, functions, statements and procedures. The "PASCAL" programming language is almost completely incorporated in Estelle and is also used for type definition. Communication between tasks is achieved via bidirectional **channels** connecting two **interaction points** of different module instances. An unbounded FIFO queue is associated with each interaction point, buffering messages (called **interactions**) at the receiving side.

A module definition in Estelle consists of a *module header* and several possible *module bodies*. The header defines the external interface of module instances and contains formal parameters, external interaction point declarations, and export/import statements allowing a restricted use of shared variables between parent and child modules. The body contains child module declarations, initialization statements, and transition descriptions defining the behavior of the module. The textual nesting of module definitions and module instance declarations defines the static *task structure* of an Estelle specification. Module instances may also be created and deleted dynamically during state transitions. Either static or dynamic task creation is followed by explicitly connecting interaction points to establish communication links between module instances. The *communication structure* obtained in this way may also be changed dynamically during state transitions.

As an example, the module header definition of an X.25 layer 3 protocol machine (X_25_DTE, depicted in Figure 15) is given in Figure 16. By substituting parameters, module instances created using this header are associated with a particular logical channel (see Figure 18 for the type definitions). Module instances receive network SPs from a network service user through the queue associated with the interaction point NSAP and send corresponding SPs back to the network service user. X.25 PLP packets are sent to and received from the X.25 network through the interaction point Packets. The corresponding channel definitions are illustrated in Figure 16. Each direction of message transfer is associated with a **role** that tasks using the channel may take.

As is apparent in the example, interactions may convey parameters. The types of parameters and of variables used in an Estelle specification are described in "PASCAL" (see Figure 18). For managing several logical channels independently and in parallel, a parent module may be defined in the specification or arrays of NSAP and Packets interaction points may be used. A parent module with several children represents a subsystem in Estelle and may itself be passive. Active and passive module instances can be attributed differently in Estelle as **process** or **activity**, **systemprocess** or **systemactivity**. Attributing system parts affects possible parallelism, called *synchronous* and *asynchronous parallelism* in Estelle. For details, see [ISO9074].

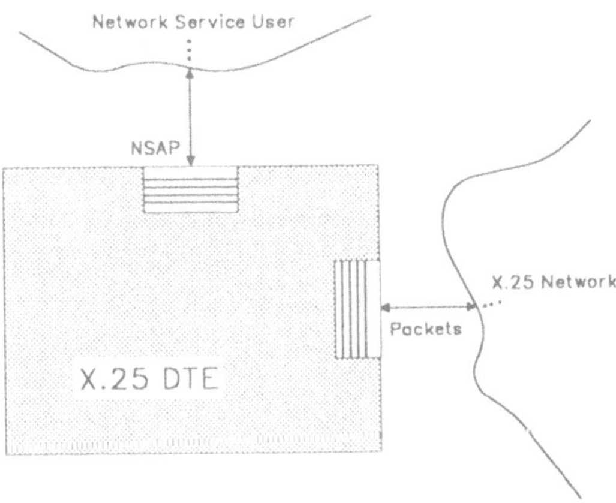

Figure 15. X.25 DTE in Estelle

In order to create a module instance, a module header is associated with a module body in an **init**-statement (see Figure 17). The interaction points of the task are connected to interaction points of other tasks by using **connect**- (for sibling tasks) or **attach**-statements (for child tasks). All module instances created run in parallel according to their attributes. During dynamic execution of the system, module instances may be deleted or communication links may be redirected by using **release**-, **disconnect**-, and **detach**-statements.

```
module X_25_DTE activity ( LC : Logical_channel_number );
   ip NSAP    : Network_service_access_point ( PROVIDER );
      Packets : X_25_network_access_point    ( DTE      );
end;

channel Network_service_access_point ( USER, PROVIDER );
   by USER     : N_CONNECT_req ( N_CR_SP : N_CONNECT_request    );
      ...
   by PROVIDER : N_CONNECT_ind ( N_CI_SP : N_CONNECT_indication );
      ...;

channel X_25_network_access_point ( DTE, DCE );
   by DTE : Call_request  ( CR_PDU : CALL_REQUEST  );
      ...
   by DCE : Incoming_call ( IC_PDU : INCOMING_CALL );
      ...;
```

Figure 16. Estelle module header and channel definitions for X.25 PLP

```
modvar dte_1 : X_25_DTE ( lc );
   ...
   initialize

      begin
         init dte_1 with X_25_DTE_body;
         ...
         connect dte_1.NSAP    to ... ;
         connect dte_1.Packets to ... ;
      end;
```

Figure 17. Creating an X_25_DTE in Estelle

```
TYPE

   N_CONNECT_request = RECORD
                          called_address       : ARRAY [1..MAX_ADD] OF CHAR;
                          calling_address      : ARRAY [1..MAX_ADD] OF CHAR;
                          receipt_confirmation_selection : BOOLEAN;
                          expedited_data_selection       : BOOLEAN;
                          quality_of_service   : Network_quality_of_service;
                          user_data            : ARRAY [1..MAX_UDAT] OF OCTET
                       END;
   ...

   CALL_REQUEST       = RECORD
                          header               : Packet_header;
                          calling_dte_address_length     : INTEGER;
                          called_dte_address_length      : INTEGER;
                          called_dte_address   : BCD_STRING;
                          calling_dte_address  : BCD_STRING;
                          facility_length      : INTEGER;
                          facility_field       : Facilities;
                          user_data            : ARRAY [1..16] OF OCTET
                       END;


   Packet_header      = RECORD
                          q_bit          : BOOLEAN;
                          d_bit          : BOOLEAN;
                          modulo         : ( modulo_8, modulo_128 );
                          lc_number      : INTEGER;
                          packet_type_id : OCTET
                       END;
   ...

   Logical_channel_number = (0..4095);

   LC_status_type         = ( free, pending, busy );
```

Figure 18. Estelle ("PASCAL") type definitions for network SPs and X.25 PLP packets

Module body specifications are separated in three parts:

- a **declaration**-part, in which constants, types, variables, procedures, channels, modules etc. are declared;
- an **initialization**-part, which is used for initializing variables, creating module instances, and connecting interaction points;
- and a **transition**-part, in which possible transitions of the extended state machine are defined.

Figure 19 illustrates how a module body for module header X_25_DTE is specified in Estelle. Main states are enumerated following the **state**-keyword. Minor state variables, types, and procedures may be declared. The initial transition executed after creation takes module instances into state p1. The dynamic behavior of module instances is defined in the transition part of a

```
body X_25_DTE_body for X_25_DTE;

   state p1, p2, p3, p4, p5, p6, p7;

   VAR   LC_status : LC_status_type;
   ...

   initialize to p1;
      begin
         LC_status := free;
         ...
      end;

   trans

      from     p1
      to       p2
      when     NSAP.N_CONNECT_req
      provided LC_status = free
      begin
         CR_PDU.header.q_bit     := FALSE;
         CR_PDU.header.d_bit     := N_CR_SP.receipt_confirmation_selection;
         CR_PDU.header.lc_number := LC;
         ...
         output Packets.Call_request ( CR_PDU );
         LC_status               := pending;
      end;
   ...
end;
```

Figure 19. Estelle module body for X_25_DTE (in part)

module body specification (following keyword **trans**). A transition from state p1 to state p2 may be performed after receiving an N_CONNECT_req interaction at interaction point NSAP. The **when**-clause in a transition specification refers to the interaction at the head of the interaction point queue. Constraints on minor state variable values and interaction parameters can be given in the **provided**-clause. Timing aspects are expressed using the **delay**-clause. All these clauses together form the *transition condition*. The state reached when performing the transition is defined with the **to**-clause. *Transition actions* are given as a list of "PASCAL" statements between **begin** and **end**. These statements describe the mapping from SP parameters to PDU fields and vice versa or modify minor state variables. An interaction is sent with an **output**-statement. The complete behavior of the extended state machine is given as a list of transition specifications (probably nested).

Protocol entities are represented as module instances in Estelle and extended state transitions describe the rules of a protocol. A service is specified as the input/output behavior of a system being composed of several communicating module instances. Layers in a communication system are represented as a hierarchy of subsystems and tasks. SPs are exchanged via channels, with interaction points modelling SAPs. PDUs may be sent and received as interactions or may be represented as structured interaction parameters. A child module can be created for each new connection.

The Estelle semantics is formally defined in [ISO9074] using transition systems. A specification can be composed of individual parts by declaring module variables and connecting their interaction points. The abstraction and expressiveness of Estelle specifications are, however, considered to be too low. This is mainly due to the employment of a programming language for "specifying" transition conditions, transition actions, and for describing all data types. The resulting specifications tend to restrict possible design choices severely and are close to implementations. Overspecification can hardly be avoided.

The architectural semantics of Estelle is also often criticized in the context of communication systems [Viss86b] [Cour87]. The fact that transitions cannot be interrupted requires a non-blocking send for output interactions. As a consequence, unbounded buffers are needed at the receiving endpoint and the communication mode is purely asynchronous. No other communication modes for exchanging interactions exist in the [ISO9074] definition, but others have been proposed in the literature because pure asynchronous communication leads to a number of problems. For example, a slow receiver is not able to impose backpressure on a fast sender; synchronized use of interaction parameters (for example, controlled release of connection endpoint identifiers) cannot be achieved; many potential "cross-overs" (messages crossing each other on the channel or waiting in the tail of the input queue) have to be handled explicitly in the specification because they may cause inconsistencies (especially for service aspects) (see also [Gotz89] [Cour87] [Viss86b]).

The distinction between synchronous and asynchronous parallelism in Estelle is of limited relevance for distributed systems. Often, the **process**-concept is not used in applications of the technique. Structuring the specification is mainly done with modules and "PASCAL" mechanisms (functions, procedures). Transitions may be nested on a textual level, but semantically the transition part is a flat list. The delay clause can be used for characterizing timing conditions of transitions, but it has no real-time meaning.

SDL

SDL is similar to Estelle in many respects, but uses a graphical syntax (SDL/GR). It is only recently that a textual syntax (SDL/PR) has been developed to ease the distribution and machine processing of SDL specifications. An SDL **system** is composed of a number of blocks that are connected via **channels**. Blocks can be further substructured into several smaller blocks, leading to a hierarchical system structure. At the lowest level, a block is defined by one or several processes, which are the parallel units. Whereas blocks define the static system structure, processes are created dynamically and are deleted after their termination. The channels connecting blocks can be refined within block substructures; the messages being conveyed are called **signals** in SDL. On the process level, **signalroutes** are attached to channels. The semantics of channels and signalroutes are different. Whereas channels cause an unspecified but perceptible delay for signals, signalroutes pass signals without any delay from source to destination. Each process has a FIFO queue attached to it that buffers the incoming signals.

A small SDL system representing an X.25 DTE by one block and one process is depicted in Figure 20. The SDL symbol for a block is a rectangle, processes are drawn as 8-cornered boxes. The block `X_25_manager` is connected to the environment via the channels `NSAP` and `Packets`, representing tho communication links to the network service user and the X.25 network. The signal lists in brackets specify what kind of signals can be sent and received via these channels. All signals have to be declared using the text symbols included in the diagram. The process `X_25_DTE` is parameterized by a logical channel number (`lc`) and has two signalroutes (`SP`, `Pack`) attached to the corresponding channels (`NSAP`, `Packets`).

Figure 20 is both an SDL *system diagram* and a *block diagram*, showing the block structure of the system and the process structure of the blocks.

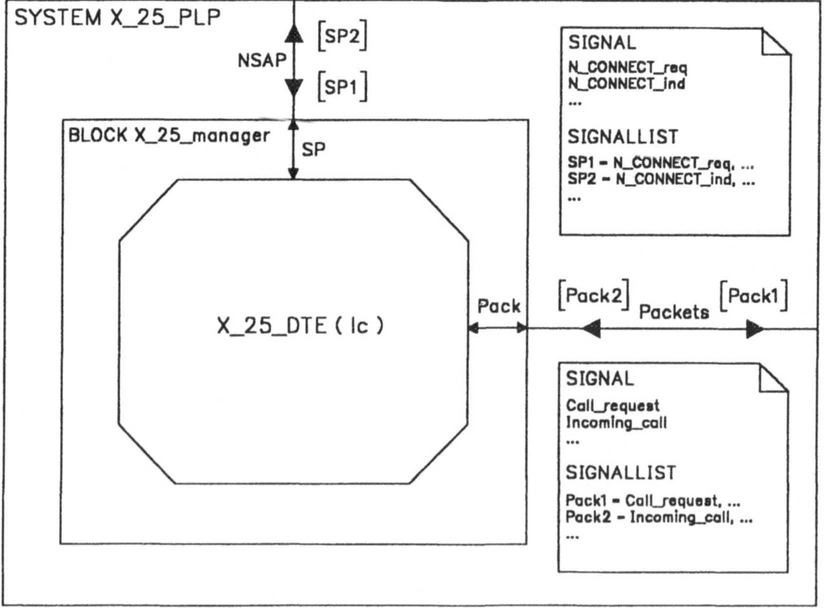

Figure 20. X.25 DTE in SDL

The process X_25_DTE has to be further refined in a *process diagram*, which is shown in Figure 21. After creation, the process X_25_DTE is in state p1. The input signal N_CONNECT_req may be received in that state provided the status of the logical channel is correct. The assignments to parameters and minor state variables are given in a task symbol. The signal Call_request is sent and a transition to the state p2 is performed. Subsequent transitions from p2, as well as alternative transitions from p1 have to be specified for the process X_25_DTE to represent the rules of the X.25 PLP.

SDL employs an algebraic approach for specifying data types of parameters and variables. Users of the technique can utilize a standard type library, containing a number of predefined types, and are also provided with syntactic abbreviations. Some examples are given in Figure 22.

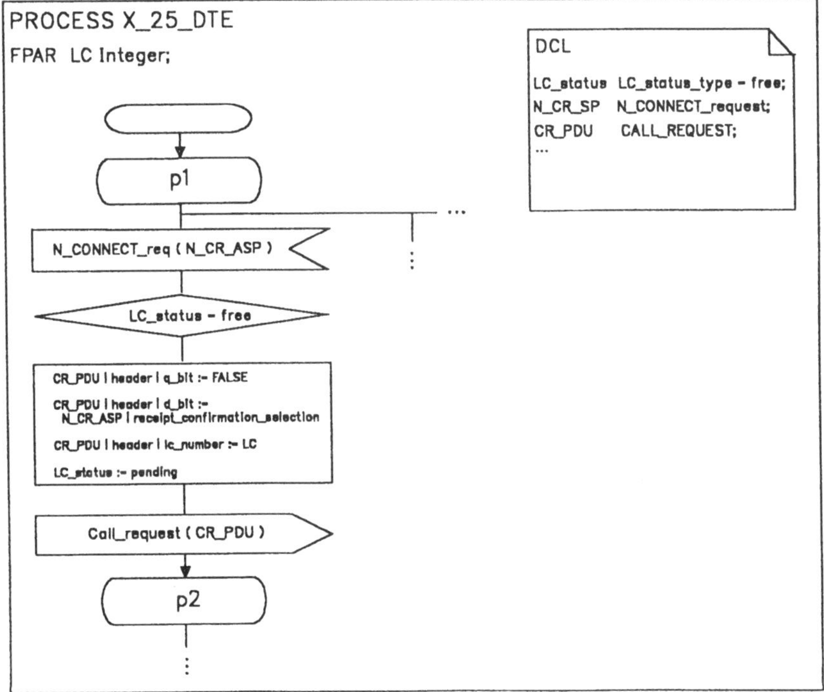

Figure 21. SDL process for X_25_DTE

```
NEWTYPE N_CONNECT_request
   STRUCT
   called_address                  : Charstring;
   calling_address                 : Charstring;
   receipt_confirmation_selection : Boolean;
   expedited_data_selection        : Boolean;
   quality_of_service              : Network_quality_of_service;
   user_data                       : Octetstring;
ENDNEWTYPE
 ...

NEWTYPE CALL_REQUEST
   STRUCT
   header                      : Packet_header;
   calling_dte_address_length : Integer;
   called_dte_address_length  : Integer;
   called_dte_address         : BCDString;
   calling_dte_address        : BCDString;
   facility_length            : Integer;
   facility_field             : Facilities;
   user_data                  : OctetString;
ENDNEWTYPE
 ...

NEWTYPE Packet_header
   STRUCT
   q_bit                      : Boolean;
   d_bit                      : Boolean;
   modulo                     : Modulotype;
   lc_number                  : Integer;
   packet_type_id             : Octet;
ENDNEWTYPE

NEWTYPE Modulotype
   LITERALS modulo_8, modulo_128;
ENDNEWTYPE

NEWTYPE LC_status_type
   LITERALS free, pending, busy;
ENDNEWTYPE
 ...
```

Figure 22. SDL type definitions for network SPs and X.25 PLP packets

One or several processes contained in one block can be used to represent a protocol entity in SDL [Beli89]. The rules of the protocol are reflected in the transitions and internal behavior of the processes. Different blocks connected via one or more channels define a service, whereas complete layers or communication subsystems are represented using block substructures. SPs are exchanged in the form of signals and parameters being conveyed may represent PDUs. A new process can be created for each connection or several signalroutes may be used. SAPs are represented as bidirectional channels, a representation which does not exactly match the intended meaning of this architectural concept.

Most of the criticism for Estelle also applies to SDL: a low level of abstraction, only asynchronous communication, limited structuring facilities for state transitions etc. In contrast to Estelle, SDL allows ADTs to be defined and makes no explicit use of a particular programming language. Other SDL characteristics are considered drawbacks compared to Estelle: only one common input queue for each process, no spontaneous transitions, deterministic selection of inputs. These characteristics complicate implementation-independent specification and appropriate extensions have been proposed for SDL in [Hogr89b].

Many of the SDL features cause controversial discussions when applied in practice. The graphical syntax is intelligible for specifications of small systems and the specifications are constructive. However, complex real-world protocols require process diagrams extending over many pages (or screens), and mixed with many text symbols. Some of the predefined constructors and generators for algebraic data type definition need thorough understanding of the underlying theory to be used properly. The SDL timer facility has considerable impact on the overall SDL semantics. On expiration, a timer signal is appended to the input queue of a process, thus interfering with the medium. With a reset operation the timer signals are removed from the input queue, regardless at what position they are, violating the FIFO property.

It is only recently that SDL has been influenced by academic research [Hogr89a] and a formal semantics defined [CCZ100]. This influence resulted in some clarifications and in the introduction of the algebraic data type concept. Unfortunately, the application of SDL for more than 20 years without having syntax and semantics formally defined has led to various SDL dialects in the past. Tool environments often use different versions of the technique.

2.2.2 Numerical Petri Nets

Petri net models have frequently been employed for describing the concurrent execution of a protocol by several entities. Starting from *condition/event nets* originally developed by Petri in 1962, a large number of different net classes have been invented and used. Condition/event nets are composed of places and transitions that are connected by directed arcs to form a bipartite graph. The places hold one token at most and tokens are simple in nature (not structured and without properties). The occurrence of transitions only depends on the presence of tokens at input places and tokens are moved to output places when the transition occurs. The modelling capabilities of condition/event nets were soon considered to be too simple for many intended application areas. The model was extended by allowing multiple tokens to exist at places and to be moved to other places during the occurrence of a transition. This led to *place/transition nets.* A further useful generalization allows for several token "colors". Individual tokens may belong to different token classes, leading to *colored nets* or *high-level Petri nets (HLPNs)* (see for example [Jens87] [Muel88]).

High-level Petri nets

A **high-level Petri net (HLPN)** (without capacity) is defined as

$N=(P, T, F, C, W, m_0)$, where

P is a set of places,

T is a set of transitions ($P \cap T = \emptyset$),

$F \subseteq (P\times T) \cup (T\times P)$ is the flow relation,

C is the data set function (or color function), assigning
a data set C_p to each $p\in P$ and
a data set (of parameters) C_t to each $t\in T$,

W is the weight function, assigning
a multiset $W(p,t)\in \mathbf{M}(C_p\times C_t)$ to each $(p,t)\in F\cap(P\times T)$ and
a multiset $W(t,p)\in \mathbf{M}(C_t\times C_p)$ to each $(t,p)\in F\cap(T\times P)$,

$m_0 \in M(p \times CP)$ is the initial marking, where

$CP := \bigcup\{ C_p \mid p \in P \}$ and $CT := \bigcup\{ C_t \mid t \in T \}$

(**M**(S) denotes the multiset of a set S).

HLPNs are usually represented as bipartite directed graphs, with expressions attached to vertices and edges (see for example Figure 23). The places are drawn as circles and boxes represent the transitions. A transition is *enabled*, if its transition condition and input condition are satisfied and tokens are located at each input place for the current marking. When the transition occurs, tokens are removed from input places and are created at output places according to the net definition; a new marking is obtained (*atomic step*).

Concurrent activities of protocol entities and inherent parallelism is modelled well by HLPNs. The rules of a protocol are represented as transitions that, during their occurrence, consume inputs and produce outputs as defined by the net structure. The possibility of using different token colors and expressing conditions on transitions and input parameters allows the

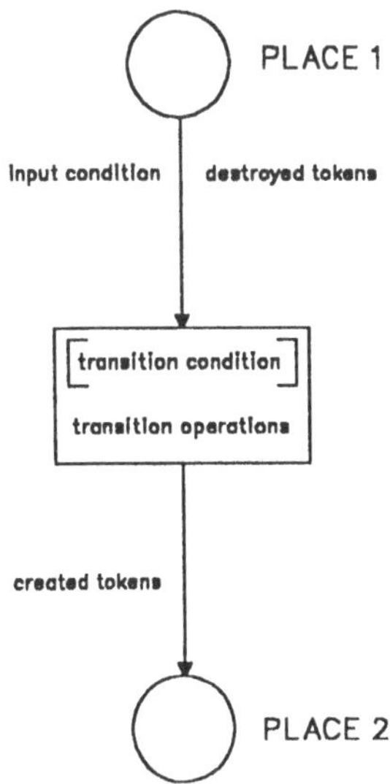

Figure 23. High-level Petri net graphical representation

specification of parameter dependencies. Powerful methods for analyzing general properties of net specifications, such as reachability, invariant conditions, deadlocks, liveness etc., are available from net theory.

NPNs

Numerical Petri Nets (NPNs) [Whee85] are a form of HLPNs being applied for specifying services and protocols [Bill88] [Diaz86] [Bill86]. The graphical representation illustrated in Figure 23 corresponds to the graphical syntax of NPNs [Bill88].

In NPNs, tokens are tuples of variables. The arcs can be inscribed with functions and expressions, leading to complex enabling and firing rules. In addition, a set of global data variables can be associated with a net specification. Only a number of simple data types are provided (Integer, Boolean, Enumerated, String). Two different capacity functions are given, restricting the number of tokens for a specific color and the total number of tokens at places, respectively.

To continue with the X.25 PLP example, there is no distinguished solution among the different possibilities for modelling an X.25 DTE by an NPN. The rules of the X.25 PLP have to be represented as transitions, SPs and PDUs are to be represented as tokens. The protocol entity is defined as a subnet of a complete net describing the behavior of the network layer. The service provided by the layer is the input/output behavior defined by transition sequences and observed as deletion and creation of tokens at particular places. In general, plenty of possibilities exist for modelling layers, services, and protocols by NPNs [Diaz86].

To further illustrate the various net mechanisms that can be used for specifying distributed systems, two small nets are shown in Figure 24, modelling different communication modes between system parts. In net (a), a shared place is used to connect the entities P1 and P2 represented by two places and one transition each. In order to communicate, one entity puts a token on the shared place and the other may remove it from there (asynchronous communication mode). In net (b), P1 and P2 are coupled by a shared transition and tokens removed from one input place may be moved to either

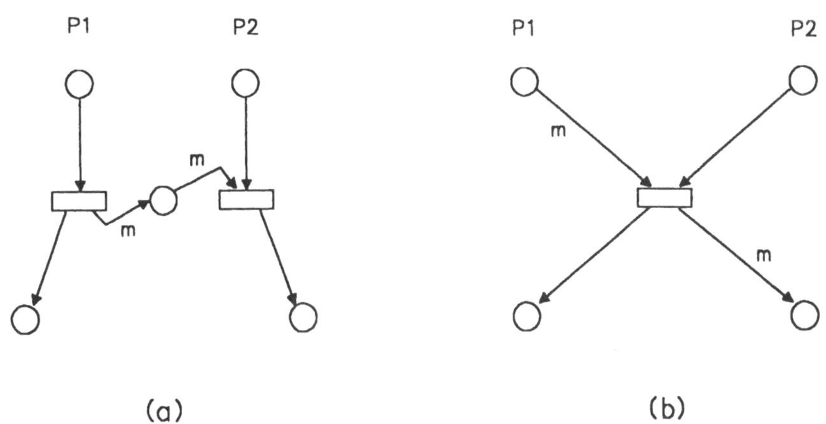

Figure 24. Modelling different communication modes with Petri nets

output place when the transition occurs (synchronous communication mode).

A transition of the NPN representing an X.25 DTE is given in Figure 25 (without capacities). The shared places NSAP and Packets are used to model the communication with the network service user and the network, respectively. A network service user puts an SP in form of a token on the NSAP place to initiate the establishment of a connection. The token is a tuple of variables, each variable represents an SP parameter. Whether the X.25 DTE is ready to accept the SP is modelled by the place SPs holding a token for each SP that can be received in a given net situation (local service constraints). The transition NConReq is enabled if tokens are present at both of the input places for an actual marking and the global variable LC_status has the correct value. When the transition occurs, a tuple of variables representing the CALL REQUEST packet is put on the place Packets and a "Call_connected" token is generated for the place PDUs. Similar to the place SPs, the place PDUs holds a token for every packet that is expected from the network in a given net situation. The global variable LC_status is set to pending.

Although the modelling power of NPNs is high, the representation of basic architectural concepts of communication systems is not straightforward.

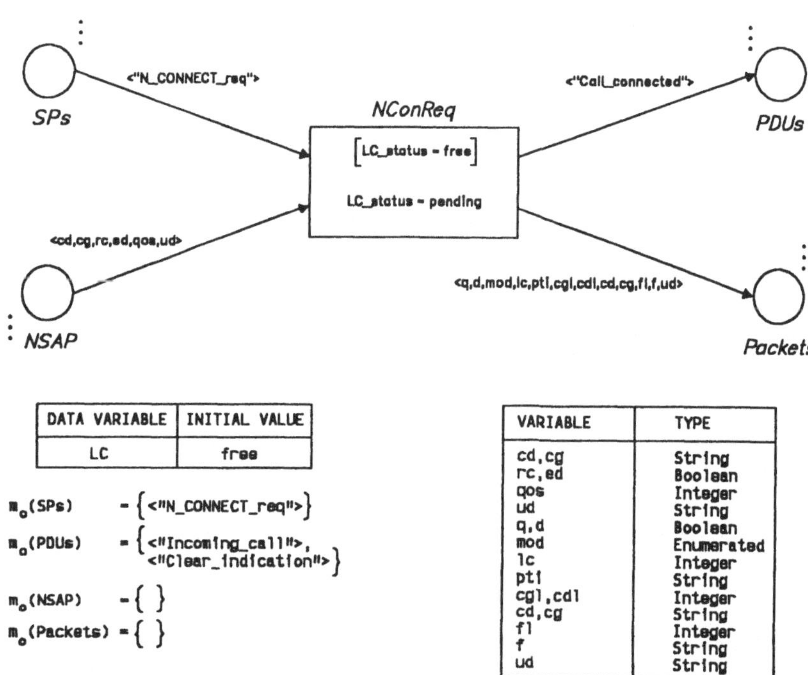

DATA VARIABLE	INITIAL VALUE
LC	free

VARIABLE	TYPE
cd,cg	String
rc,ed	Boolean
qos	Integer
ud	String
q,d	Boolean
mod	Enumerated
lc	Integer
pti	String
cgl,cdl	Integer
cd,cg	String
fl	Integer
f	String
ud	String

Figure 25. X.25 DTE as Numerical Petri Net

First, basic concepts like entities or SAPs require several places and/or transitions to be represented, due to the very general mechanisms that nets offer. Second, the variety of adequate possibilities for describing a system can cause confusion. Each new net has to be understood from the beginning by its low-level primitives because abstraction mechanisms are missing.

HLPNs (and NPNs) are formally well-defined and powerful methods exist for analyzing nets. Compared to normal place/transition nets, the expressiveness of NPNs is better, but certain requirements can sometimes only be described by larger nets with several transitions and places. Data typing facilities are severely limited. On the other hand, nets are implementation-independent specifications expressing concurrency and causality well. Compositionality can be achieved in various ways: by identifying places and/or transitions of individually developed nets or by introducing subnets connecting components (see for example [Vogl89]). However, the latter strategy may change the overall meaning of the net specification considerably. Net specifications are interpreted in terms of their transition system semantics and are constructive in nature. The intelligibility of large nets is, however, considered critical because of the complex dynamic behavior.

2.2.3 LOTOS

The formal description technique LOTOS (*Language of Temporal Ordering Specification*) was developed within ISO between 1981 and 1988 [ISO8807] (in parallel with Estelle). A *process algebraic* approach was taken and the algebraic data type language ACT ONE was adopted for abstract data type definition. A comprehensive tutorial on LOTOS is provided in [Bolo87].

Process Algebras

The process algebraic approach to the specification of communicating systems was initiated by Milner with *CCS (Calculus of Communicating Systems)* presented in [Miln80]. The basic idea is to describe a system only by the behavior that can be seen (or experienced) by an external observer. Systems are composed of several communicating *agents* interacting with each other and with the external environment. Communication takes place as an indivisible unit of synchronized interaction (called *actions*). The behavior of agents is described by *behavior expressions* that are subject to algebraic laws. Several related approaches (for example [Hoar85]) use the notion of *processes* instead of agents.

Agents are often graphically represented as boxes, with the set of possible actions drawn at the sides (see Figure 26 for an example). The observable actions determine the alphabet Λ of an agent. The unobservable (internal) actions of an agent (for example, internal communication between two components) are represented with an internal action symbol. The behavior of an agent is specified by an associated behavior expression (see Figure 26) that is constructed from predefined simple behavior by applying a fixed set of compositional operators. These operators can be used to combine agents sequentially or concurrently and to specify alternative or recursive behavior. The number and characteristics of predefined simple behaviors and available operators distinguish different process algebras. The syntax employed is also different in many cases.

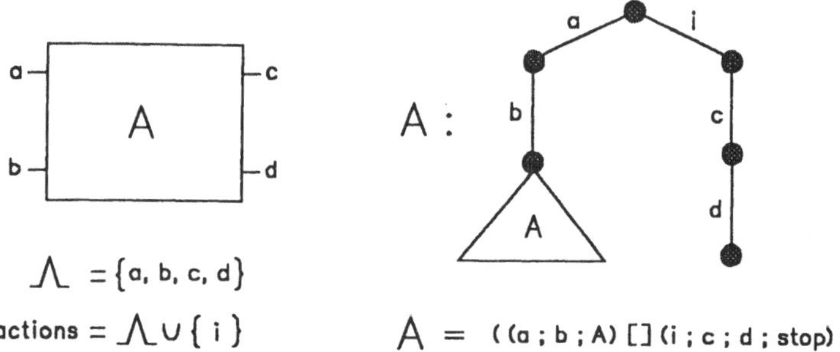

Figure 26. Representation of agents in a process algebra

The behavior of an agent is often visualized by an *action tree* (see Figure 26). In an action tree, edges are labelled by actions. Nodes correspond to initial or intermediate behavior expressions, leading to an implicit notion of "states" (node labels have been omitted). The example agent A in Figure 26 is able to perform an arbitrary number of 'a' and 'b' actions in sequential order. Alternatively, the agent may decide to perform an internal action, participate in actions 'c' and 'd', and remain inactive forever. This behavior is represented as a behavior expression given underneath the action tree in Figure 26. The syntax of LOTOS has been employed as summarized in Figure 27. When an agent is composed in parallel with other agents, their common actions have to be performed synchronously (they observe each other). All other actions may take place in arbitrary order, only confined by the order given for the individual agents. In this way, concurrency is modelled by arbitrary interleaving.

The semantics of operators and behavior expressions in a process algebra are usually given by a set of axioms and inference rules. The inference rules determine how behavior expressions transform into others after actions have been performed and communication has taken place. Strong algebraic laws can be postulated for the operators and are the basis of powerful process algebraic calculi. The notion of *observational equivalence* is of particular importance for verifying specification-implementation relations between agents.

In protocol specification, actions are associated with SPs or PDUs and consist of a SAP identifier, identification of the SP or PDU, and parameter values. Ordering constraints are defined by the behavior expressions of agents modelling protocol entities and service providers. The execution of particular actions may depend on parameter values. Parallel composition of agents yields a layer or a complete communication system.

LOTOS

In LOTOS, communicating agents forming a system are called **processes**. A specification defines a hierarchy of processes, where process behavior is expressed by combining subprocesses exhibiting simpler behavior until the specification can be expressed directly (*inaction, action prefix, choice, successful termination*). Composition operators are provided for sequential composition (*enabling, disabling*) or parallel composition of processes (*pure interleaving, full synchronization, general case*). Processes being composed in parallel communicate by participating in a synchronous event. The notion of a **gate** is used as a synonym for event labels and symbolizes the abstract location where the interaction takes place. Interactions performed at internal gates can be hidden from the outside world by using the *hiding*-operator. The symbol **i** is used to denote an internal event. Existing process definitions can be instantiated in other process definitions. In this way, recursive behavior can be achieved.

The syntax employed in Basic LOTOS is summarized in Figure 27 [Bolo87]. Basic LOTOS is the subset of the language dealing with a finite set of possible events to express process behavior, without using data parameters. In Full LOTOS, data parameters are associated with events and processes may communicate data values during an interaction. Data values are also used for instantiating parameterized processes, defining generalized choices and guarded expressions, or passing information to subsequent processes.

In the following example, g is a gate, E is a value expression, and B1 and B2 are processes. Value communication between B1 and B2 may take three different forms in Full LOTOS:

```
                         B1             B2
(1) value matching   :   g ! E1         g ! E2        value(E1) = value(E2)

(2) value passing    :   g ! E          g ? x : t     value(E) of sort t

(3) value generation :   g ? x : t      g ? y : u     t = u
```

Since several processes can be composed in parallel at one gate and since these processes have to synchronize their actions (*multi-way synchronization*), value communication can occur between any number of processes. In case (1), synchronization is achieved if the values offered by all processes are equal. In case (2), the value offered is passed to other processes declaring variables for receiving that value. Since no explicit value is offered in case (3), a value being suitable for all participating processes is generated and received by all processes. The set of suitable values may be restricted by a boolean expression (*selection predicate*). A high level of abstraction can be achieved by writing the formal description in such a

Name	**Syntax**
inaction	**stop**
action prefix	
- unobservable (internal)	**i;B**
- observable	**g;B**
choice	**B1 [] B2**
parallel composition	
- general case	**B1 |[g_1, ..., g_n]| B2**
- pure interleaving	**B1 ||| B2**
- full synchronization	**B1 || B2**
hiding	**hide g_1, ..., g_n in B**
process instantiation	**p [g_1, ..., g_n]**
successful termination	**exit**
sequential composition (enabling)	**B1 >> B2**
disabling	**B1 [> B2**

Figure 27. Syntax of behavior expressions in Basic LOTOS

way that several processes acting in parallel impose their individual constraints on data values during synchronization with other processes. This is called the *constraint-oriented* specification style. Other specification styles exist and should be used for more implementation-oriented specifications.

A small excerpt from a LOTOS specification of an X.25 DTE is given in Figure 29. The process X_25_DTE is composed in parallel with other processes describing the behavior of the network service user and the X.25 network (together forming the process X_25_manager). The process configuration is illustrated in Figure 28, where the process X_25_DTE is drawn as a box. The gates where interactions with the external environment may occur are drawn at the sides of the box (NSAP, Packets). The parameters of the process X_25_DTE are the synchronization gate NSAP (for communication with the service user), the synchronization gate Packets (for communication with the network), and the logical channel number lc. The behavior of the process X_25_DTE is defined locally in the process X_25_manager, as indicated by the *where*-clause. Three sequentially composed subprocesses are used, corresponding to the three phases of the X.25 PLP. Successful termination of the connection establishment phase (process ConEst) enables data transfer (process DatTra). Either phase may be disrupted by connection release (process ConRel).

The process ConEst is able to participate in an interaction at the gate sp provided that the first value offered in the event expression matches "N_CONNECT_req" and the remaining values for the variable N_CR_SP are passed. Subsequently, an interaction at the gate pack is performed and the values of the CALL REQUEST packet are passed to the network. Subsequent behavior according to the X.25 PLP is to be specified using action prefixing, or alternative behavior can be defined using the choice-operator.

As can be seen in the example, values offered at gates can be complex expressions composed of variables, constants, and operation symbols. The variables receiving the value of those expressions must have the same type and the semantics of the operations must be defined in order to allow an evaluation. The types and operations are specified with ACT ONE, as illustrated in Figure 30. Abstract data types are defined algebraically by defining a signature (sort symbols, operation symbols, and their domain

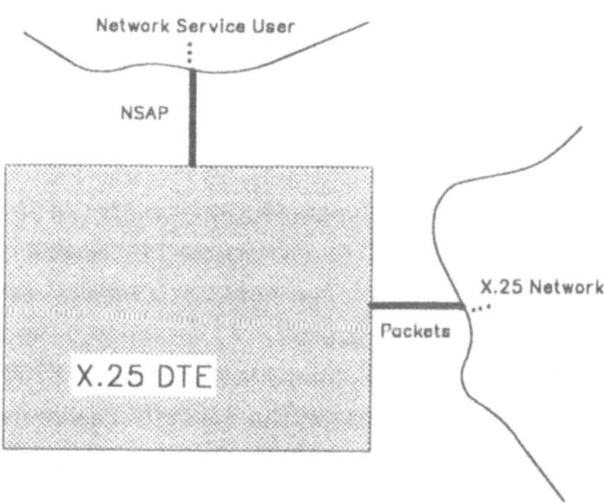

Figure 28. X.25 DTE in LOTOS

and range) and giving the semantics of operations by a set of equations which relate terms (see for example [Ehri85]).

```
process X_25_manager [NSAP, Packets] ( lc : Nat ) : noexit :=
 Network_Service_User | [NSAP] |
    X_25_DTE [NSAP, Packets] ( lc ) | [Packets] |
       X_25_Network [Packets]
 where

 process X_25_DTE [SP, Pack] ( LC : Nat ) : noexit :=
  ( let LC_status : LC_status_type = free in
   ( ( ConEst [SP, Pack] ( LC ) >> DatTra [SP, Pack] ( LC ) )
     [> ConRel [SP, Pack] ( LC ) )
  where

  process ConEst [sp, pack] ( LC : Nat ) : exit :=
   [LC_status = free] ->
    (sp !"N_CONNECT_req" ? N_CR_SP : N_CONNECT_request ;
     pack !"Call_request
          ! CR_PDU ( make_header ( false,
                                   receipt_confirmation_selection ( N_CR_SP ),
                                   modulo_8,
                                   LC,
                                   CR_pti ),
                                   ... ); ...
   [] ...
  endproc ...
 endproc ...
endproc
```

Figure 29. LOTOS process for X_25_DTE

```
type N_CON_req is Boolean, NaturalNumber, CharString, OctetString,
                  NetworkQualityOfService
 sorts N_CONNECT_request
 opns
  NConReq : CharString, CharString, Bool, Bool, NetworkQoS, OctetString ->
            N_CONNECT_request
  receipt_confirmation_selection : N_CONNECT_request -> Bool
  ...
 eqns
  forall cd, cg : CharString, rc, ed : Bool, ud : OctetString
   ofsort Bool
    receipt_confirmation_selection ( NConReq ( cd, cg, rc, ed, ud ) ) = rc ;
    ...
endtype
 ...

type Packet_header is Boolean, Modulo, NaturalNumber, Octet
 sorts Header
 opns
  make_header : Bool, Bool, Mod, Nat, Octet -> Header
  q_bit       : Header -> Bool
  ...
 eqns
  forall b1, b2 : Bool, mod : Mod, no : Nat, id : Octet
   ofsort Bool
   q_bit ( make_header ( b1, b2, mod, no, id ) ) = b1 ;
   ...
endtype
 ...

type CALL_REQ is Packet_header, NaturalNumber, BCDString, Facilities,
                 OctetString, Octet
 sorts CALL_REQUEST
 opns
  CR_PDU : Header, Nat, Nat, BCDString, BCDString, Nat, Facs, OctetString ->
           CALL_REQUEST
  CR_pti :                 -> Octet
  header : CALL_REQUEST -> Header
  ...
 eqns
  forall cr : CALL_REQUEST, h : Header, cgl, cdl : Nat, cd, cg : BCDString,
         fl : Nat, f : Facs, ud : OctetString
   ofsort Octet
   CR_pti = '0B'H
   ofsort Header
    header ( CR_PDU ( h, cgl, cdl, cd, cg, fl, f, ud ) ) = h ;
   ...
endtype
 ...
```

Figure 30. LOTOS (ACT ONE) type definitions for network SPs and X.25 PLP packets

Since LOTOS was conceived for specifying OSI services and protocols, the architectural concepts of communication systems are represented well. Processes model protocol entities and can be composed flexibly to form a protocol layer and to provide a service. SPs and PDUs are exchanged with interactions at gates representing SAPs. The functionality of a service or the rules of a protocol are defined by temporal orderings of interactions and multi-way synchronization constraints. A new process can be instantiated for each new connection and processes can be parameterized by CEPs.

Formal descriptions in LOTOS are constructive specifications that are interpreted in terms of *labelled transition systems* semantics [ISO8807]. An elegant and powerful theory exists for equivalences between processes, equivalence-preserving transformations, and testing (see [Eijk89a]). Several operators are provided for parallel and sequential composition of processes to form complex systems. The degree of abstraction is high because the focus is on external observability and nothing is stated on internal details of processes.

Nevertheless, creating LOTOS specifications is considered to be a complicated task. Because of lack of other structuring mechanisms, the process concept is applied at different levels of detail. This leads to highly recursive behavior expressions that are difficult to read. The missing concept of an explicit internal state can force specifiers to describe parameter mappings between interactions by awkward constructions (*let, accept*). Scope rules for variables introduced in event expressions are hard to recognize in the specification text. The same applies for priority rules of operators.

The concept of multi-way synchronization is considered inadequate for distributed system design because it will rarely be implemented in practice. Creating specifications according to different specification styles is part of the LOTOS design methodology, but is not really enforced by the technique. Mixed-style specifications can easily be produced, although that was never intended.

The application of ACT ONE for specifying data types requires a substantial effort. Only limited support is provided by a predefined type library and all other types have to be defined from scratch. The denotation of values as terms composed of variables, constants, and operation symbols is not practical. Some predefined abbreviations have to be provided to users of the technique.

2.2.4 Z

The Z notation is a general-purpose formal description technique developed at Oxford University since 1980. It is based on set theory and predicate logic and has been employed for specifying a variety of information processing systems [Haye87]. More recently, Z has also been used in the area of communication systems [Duke88]. An introduction and reference manual is provided in [Spiv89].

Predicate Logic

A language of formal logic is defined in two steps (see for example [Mann74]). First, syntactic rules are given that determine the set of well-formed formulas. In **first order predicate logic (FOPL)**, symbols exist for constants, individual variables, functions, predicates, quantors, and logical connectives ($\wedge$, $\vee$, $\neg$, $\Rightarrow$, $\Leftrightarrow$, $\forall$, $\exists$). *Terms* are defined as constants, variables, or function symbols applied to terms. An *atomic formula* is either a predicate symbol applied to terms, or an equality between two terms. A *formula* can be an atomic formula, or may consist of formulas combined by logical connectives.

In a second step, semantics of well-formed formulas are defined. This is done by introducing *interpretations*, which are mappings from the variable symbols to values from a specific domain of individuals, from constant symbols to constants, and from function and predicate symbols to functions and predicates over the selected domain. Every well-formed formula possesses a truth value 'true' or 'false' under a specific interpretation. Alternatively, universally valid formulas can be derived in a deduction system that consists of a language of formal logic, together with a set of logical axioms and inference rules. The fact that a formula is universally valid, or valid under a specific interpretation, can be proven by syntactic derivation in a deduction system. The formulas that are logical consequences of a given set of axioms and inference rules can be enumerated and form a logic theory.

Specification techniques based on logic are often called *descriptive* techniques, as opposed to *constructive* techniques, because logic specifications just define desired properties of a system and are not interpreted algorithmically (see the discussion on FDT requirements at the beginning of this chapter). However, algorithmic (procedural) interpretations exist for certain logic languages (for example, Horn clause logic [Kowa79]). In general, descriptive specifications are better suited for verifying system properties (see for example [Krum84]), whereas constructive specifications are better suited for implementing systems.

Z

A Z specification comprises definitions of mathematical objects like types, functions, relations, sets etc., and a number of **schemas**. Data typing mechanisms are based on sets. Several basic sets are predefined and new types can be constructed by using set operations, functions or relations over sets, cartesian products etc. Schemas provide a framework for declaring a number of objects that are associated with variables and a number of predicates that describe properties of the schema by relating the variables.

Different parts of a communication system can be represented as **state schemas** in Z. In Figure 31, an X.25 DTE is represented by the conjunction of schemas `NSAP`, `Packets`, and `DTE_state`. The schemas `NSAP` and `Packets` model communication channels by declaring variables (`sps`, `pack`) which hold sequences of messages of an underlying base type. Predefined operations, such as `head` and `tail`, are used to manipulate these variables. Sending is done by appending interactions at the end of a sequence and messages are received by removing interactions from the head of a sequence. The communication achieved this way is asynchronous. The schema `DTE_state` represents the state of the protocol entity as a number of variables and schema `Init_DTE` describes the initial values for a specific X.25 DTE.

The rules of a protocol are captured by **operation schemas** defining possible actions on state schemas. The operation schema `NConReq` shown in Figure 32 describes how establishment of a connection is initiated according to the X.25 PLP. The upper part of the schema contains the declarations and the lower part (below the separation line) consists of conditions

```
NSAP
sps : seq SP

Packets
pack : seq PDU

DTE_state
LC        : 0..4095
LC_status : {free, pending, busy}
p_state   : {p1, p2, p3, p4, p5, p6, p7}
...

X_25_DTE ≅ NSAP ∧ DTE_state ∧ Packets

Init_DTE [lc]
X_25_DTE
----------------
LC        = lc
LC_status = free
p_state   = p1
...
```

(Symbol "≅" means "is defined to be")

Figure 31. X.25 DTE in Z

```
NConReq
Δ X_25_DTE

CR_PDU  : CALL_REQUEST
N_CR_SP : N_CONNECT_request
----------------
N_CR_SP    = head ( sps' )
p_state'   = p1
LC_status' = free
...
sps                       = tail ( sps' )
CR_PDU.header.q_bit       = false
CR_PDU.header.d_bit       = N_CR_SP.receipt_confirmation_selection
CR_PDU.header.lc_number   = LC
...
p_state                   = p2
LC_status                 = pending
...
pack                      = pack' ^ <CR_PDU>
```

(Symbol "^" means "concatenation" and "<>" means "make sequence")

Figure 32. Z operation schema for X_25_DTE

and actions. Previously defined schemas can be referenced in declarations, as done for X_25_DTE, which means that all objects defined in the referenced schema are available in the new schema (*schema inclusion*). Additionally, all properties of the referenced schemas (described by predicates) are $\wedge$-connected with the predicates defined in the new schema. If the schema inclusion is preceded by the Δ symbol, the values of the included objects can be modified by actions. These actions are described by predicates defining a relation between the before-state (primed variables) and the after-state (unprimed variables). An SP sent by the network service user is retrieved from the variable sps with the head-operation, provided that the variables p_state and LC_status have appropriate values. The contents of the CALL REQUEST packet (CR_PDU) and new values of state variables are described. A new message for the network is constructed and appended to the variable pack. State schemas may be used to define types, as illustrated in Figure 33. Elements of other schemas can be accessed using a dot notation (see for example schema NConReq). SP parameters and PDU fields may also be defined by cartesian products and accessed via projection functions.

In Z, concurrency is modelled by allowing operations to be performed concurrently provided they act independently on the system state. However, this may be difficult to predict for specifiers and it may lead to unexpected inconsistencies. Z specifications are descriptive because they state only properties of systems and the usage of predicate logic abstracts from their implementation. There is, however, a constructive interpretation based on the execution history [Duke88]. Any history of the specified system starts in an initial state and from then on, enabled operations are selected and performed until forever or until no further enabled operation exists. Execution histories can be stored in schemas and may serve for verifying safety and liveness properties.

The theory of Z is well-founded in traditional mathematics. The expressiveness and the degree of abstraction is high. However, the fact that Z was not especially conceived for specifying communicating systems complicates its application in this area. This is mainly due to missing concepts for explicitly modelling various communication modes and communicating entities. Variables holding message streams are considered inadequate for representing

```
N_CONNECT_request
called_address                     : ADDRESS
calling_address                    : ADDRESS
receipt_confirmation_selection     : BOOLEAN
expedited_data_selection           : BOOLEAN
quality_of_service                 : NetworkQualityOfService
user_data                          : OCTETSTRING
...
SP ≅ N_CONNECT_req ∨ N_CONNECT_indication ∨ ...

Packet_header
q_bit          : BOOLEAN
d_bit          : BOOLEAN
modulo         : {modulo_8, modulo 128}
lc_number      : 0..4095
packet_type_id : OCTET
...

CALL_REQUEST
header                      : Packet_header
calling_dte_address_length  : ℤ
called_dte_address_length   : ℤ
called_dte_address          : BCDString
calling_dte_address         : BCDString
facility_length             : ℤ
facility_field              : Facilities
user_data                   : OCTETSTRING
...
PDU ≅ CALL_REQUEST ∨ INCOMING_CALL ∨ ...

(Symbol "≅" means "is defined to be")
```

Figure 33. Z type definitions for network SPs and X.25 PLP packets

SAPs. Complex systems can be composed of individual parts, but internal and external behavior is not differentiated within the semantics of Z. To circumvent that problem, users of the technique defined separate schemas and refer to them informally as "the external view" and the "internal view" [Haye90]. The intelligibility of Z specifications decreases the more predefined and user-defined mathematical symbols are used. Often profound mathematical knowledge is required to understand the specification. The question arises as to whether this can be expected from implementors.

2.3 Comparative Assessment

The main problems encountered with specifications created by means of natural language text and pictures are that they tend to be ambiguous, are often incomplete or imprecise, and cannot be processed by automated tools. These are severe restrictions that have to be overcome. Precise specifications that can be processed by tools are facilitated by FDTs. However, the acceptance and application of FDTs in industrial projects is still limited. While the benefits of using FDTs for the development of new communication protocols and services are gaining recognition, existing FDTs still have considerable weaknesses.

The degree of abstraction reached with Estelle and SDL is too low, resulting in overspecification. No programming language should be employed for specifying abstract services and data types. Nondeterminism as a means of expressing alternative behavior is important and should be available. Synchronous communication is appropriate for specifying abstract service aspects and should be provided as well. FDTs employing state transition models should offer language constructs that allow thorough structuring of the transition space. True parallelism within a state machine cannot be expressed.

Petri nets have proved to be an excellent formalism for modelling concurrency and causality in concurrent systems. Net specifications of concurrent systems can be analyzed in various ways for checking properties. However, the description primitives provided by Petri nets are considered inadequate for directly representing basic architectural concepts of communication systems, such as SAPs, entities, or layers. The invention of HLPNs facilitates more compressed specifications, but still does not offer the appropriate concepts. Adequate data typing mechanisms and a clear concept of modularization are not provided.

The degree of abstraction and the sound mathematical basis of the Z notation are appealing, but similar arguments apply as for Petri nets. The representation and parallel composition of protocol entities is difficult to describe. Modelling communication via SPs at SAPs according to the basic concepts of open systems is not straightforward. This is because the basic architectural concepts of communication systems have to be modelled by users of the technique using several schemas before adequate representations of these concepts are available. The notion of external observability is not reflected in the semantics of Z. Specifications are descriptive and better suited for formal verification of properties, rather than supporting a constructive approach. Recent proposals have been made to improve the notation by introducing object-oriented ideas [Duke90].

From the languages evaluated in the last section, LOTOS meets best the requirements for an FDT to be applied for specifying communication systems. The technique focuses on external observability and provides powerful concepts supporting communication system design. The abstraction from internal states of specified systems is advantageous when abstracting from particular implementations, but also poses considerable problems. Parameter dependencies, local and global constraints, and protocol rules are difficult to formalize without the concept of an internal state if they are of realistic complexity. Multi-way synchronization is considered inadequate for describing communication between distributed processes. Although providing an excellent semantical basis, the algebraic style of writing abstract data type specifications has to be improved for LOTOS in order to be applicable in practice.

The results of the evaluation of existing FDTs has been the motivation to search for an alternative approach that tries to combine the advantages of the existing techniques while avoiding their main drawbacks. The new approach is presented in the next chapter.

3 Protocol Specification with Communicating Rule Systems

A rule-based approach to the formal description of communication services and protocols is presented in this chapter. The technique is called *Communicating Rule Systems (CRS)* and can be applied to the specification of concurrent, distributed systems. Rules are employed to model the behavior of communicating system components. The basic ideas of the technique were introduced in [Mack87] and a first language definition was given in [Neum88].

CRS was conceived to combine advantages of existing techniques (see [Mack87]) and is based on several of the concepts explained in chapter 2. The approach is extensional, focussing on the behavior that can be seen by an external observer, but an extended finite state machine (EFSM) model is used to describe possible behavior sequences. The state space is given by a number of objects that can be defined using data abstraction. A formal framework based on first order predicate logic (FOPL) and employing a rule paradigm is provided to define and structure the transition space.

In this thesis, the focus is on the application of CRS to the development of communication systems, whereas in [Neum88] CRS has been presented as a technique to describe the parallel execution and communication of knowledge-based systems. Applying CRS to the area of communication systems is appealing for several reasons explained in the following sections. Many authors like to characterize communication protocols as sets of rules (see for example [Krum84] [Nehm85] [Slom87] [Tane88] etc.). The experiences gained from applying CRS to several real protocols resulted in a number of modifications and extensions to the language definition given in [Neum88]. One major change was to adopt the *Abstract Syntax Notation One (ASN.1)* standardized by ISO and CCITT as a substitute for the former CRS type concept. This was done because ASN.1 meets the requirements of a data description technique for communication systems. In order to achieve the integration into CRS several extensions to ASN.1 had to be defined concerning operations on data objects, resulting in the *Extended*

Abstract Syntax Notation One (X-ASN.1). The mechanisms defined in [Neum88] to structure the set of rules in CRS specifications have been modified and extended as well. Finally, an operational semantics based on *labelled transition systems* has been developed.

The resulting CRS language as used in this thesis is presented in the following. The data definition constructs provided by X-ASN.1 are explained in section 3.1. How rules and rule systems can be specified is described in section 3.2, including an assessment according to the criteria developed in chapter 2. The operational semantics is given in section 3.3. A number of examples taken from the X.25 packet level protocol (PLP) [ISO8208] [ISO8878] are provided as illustrations of the various concepts introduced in this chapter.

3.1 Extended Abstract Syntax Notation One (X-ASN.1)

The *Abstract Syntax Notation One (ASN.1)* evolved from the CCITT recommendation X.409 as part of the application layer standard for message handling systems to an international standard of its own. The development of the language has been aligned between the ISO and CCITT standardization bodies. Advanced tutorials on ASN.1 are provided in [Gora87] [Chap89][Gora90].

In order to achieve correct operation of a protocol, the participating protocol entities need precise information about the data structures to be exchanged. Since protocol entities are located in physically separate systems and are realized in different hard- and software environments they have to agree on the structure of the data they want to exchange in a way independent of their local representation. Following the OSI model, the nature of the data changes from simple byte sequences in the lower layers to complex structured objects in the upper layers. For describing the abstract structure of data objects exchanged by application processes an appropriate data description language is needed.

ASN.1 can be used to describe the structure of data independently of the local representations, programming languages, and implementation strategies employed for the different protocol entities. ASN.1 specifications just define the **abstract syntax** of data objects; various forms of representing these objects in different systems are possible (**local syntax**). The syntax used for transmitting the data is called **transfer syntax** and is derived from the abstract syntax by applying a set of **encoding rules**. Different sets of encoding rules may exist for different protocols. Currently, ISO has standardized the so-called *basic encoding rules* for ASN.1 in [ISO8825]. The conversion from local syntax to transfer syntax and vice versa is called **encoding/decoding** and is done by corresponding routines in each system, as illustrated in Figure 34 [Gora88b]. System A and System B in Figure 34 are using local implementations of a data type described in ASN.1, employing "MODULA" and "C", respectively, as implementation languages.

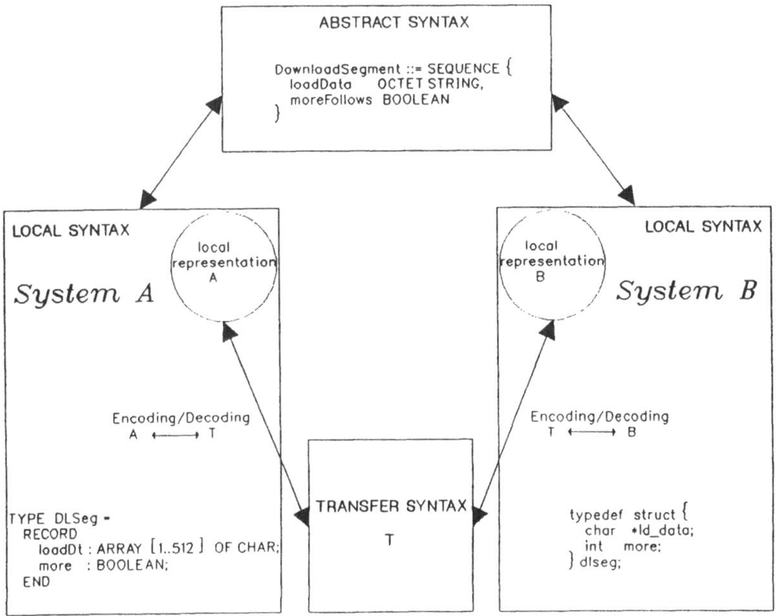

Figure 34. Abstract, local and transfer syntax in communication systems

ASN.1 was intended to be used for the implementation-independent description of application layer PDUs. However, the notation has much broader scope. The abstract structure of lower layer PDUs can be described as well (different encoding rules may have to be used) and the ASN.1 notation may also be applied to defining the structure of SPs and SP parameters [Schn90c]. The important aspects to be described for PDUs and SPs are the ordering of data, presence or absence of parameters, identification of fields, and the encoding/decoding of values. This information is crucial to the participating parties and difficult to express using common specification or programming languages. The constructs provided by ASN.1 have been conceived for specifying the structure of data to be exchanged in communication systems. They allow the definition of optional fields, different orderings, various string types, named numbers and bits, and many other useful structures. However, the imbedded use of ASN.1 in specification or programming languages was not intended by the standards committees. Consequently, the ASN.1 standard defines no means for accessing and manipulating different components of ASN.1 typed objects. The ASN.1 extensions considered for integrating the notation into CRS (X-ASN.1) are explained together with the standard ASN.1 constructs in the following.

3.1.1 Type Definition and Operations

Like most type concepts available in programming languages, ASN.1 provides a basic set of predefined types, called **simple types**, and a set of mechanisms for constructing **structured types**. The simple types include boolean values, integer and floating point numbers, and various forms of strings. Type constructors exist for defining ordered or unordered sequences and sets of values from previously defined types. Another constructor allows the definition of types as a choice of several existing types. New types may also be built as **subtypes** of existing types by restricting ranges or enumerating possible values.

ASN.1 type and value definitions have to be encapsulated in a type **module**. The syntactic framework is illustrated in Figure 35. The types and values that can be referenced in other modules are given in the export list and the types and values being used from other modules are given in the import list. The requirement to put type and value definitions into a module enforces the modular construction of data structures for communication pro-

```
X_25_Packets DEFINITIONS :=

   BEGIN

   EXPORTS ... ;
   IMPORTS ... ;
      ...
   END
```

Figure 35. ASN.1 module definition frame example

tocols. Several modules can be created for different protocol layers, service interfaces, functional units, or other aspects of grouping.

The notion of a *type* is used in ASN.1 to summarize a set of values. The ASN.1 value notation defines how values of different types are denoted. No operations are associated with ASN.1 types in contrast to *data types* in programming or specification languages. Adding operations to each of the ASN.1 types and type constructors is the intention of X-ASN.1. Thus, the ASN.1 constructs as described in [ISO8824] are fully supported by X-ASN.1. In addition, a number of operations is defined for each type that can be specified with ASN.1. In the following, the abbreviation X-ASN.1 is used when referring to the resulting data definition language as a whole.

When talking about data types where the representation of values and the implementation of associated operations is fixed, the notion *concrete data types* is often used. Data types that are defined without using a specific representation and implementation are called *abstract data types (ADTs)* (see for example [Clea86]). The basic idea of ADTs is to separate the use of a type from its implementation. By defining possible operations on ASN.1 types, predefined ADTs are provided to users of X-ASN.1. For the type constructors this results in a generic type concept. Completely new ADTs may be defined (if required) using data abstraction in CRS (see section 3.2).

Nothing has been said so far about another ASN.1 facility: the **macro notation**. By using macros, a user of the technique is able to extend the predefined ASN.1 syntax and to introduce completely new type definition constructs. Thus, the ASN.1 macro notation means much more than the textual substitution available in assembler languages. Since the syntax of ASN.1 is not fixed if macros are used, analysis tools for the language are very hard to implement. No mechanism exists to define the meaning of new syntactic constructs introduced by ASN.1 macros. The ASN.1 macro notation has been criticized by many authors (for example [Gora88a] [Neuf90]) and will probably be reconsidered by the standardization bodies. In this thesis, the ASN.1 macro facility is not used.

The simple types, structured types, and subtypes that can be defined using X-ASN.1 and the possible operations available are explained in the following. Some type definitions for data structures used in the X.25 PLP are given as examples.

Simple types

The following simple types are provided by X-ASN.1:

- BOOLEAN

  values TRUE, FALSE

  operations $\neg$ (negation), $\wedge$ (logical-AND), $\vee$ (logical-OR), $\Rightarrow$ (implication), $\Leftrightarrow$ (equivalence), IF-THEN-FI and IF-THEN-ELSE-FI (conditionals)

  comments The usual logic operations are provided, the conditional operations are considered to be useful as well.

  examples Figure 36 (1);

- INTEGER

  values all integer numbers

  operations $+_{int}$ (addition), $-_{int}$ (subtraction), $*_{int}$ (multiplication), DIV (integer division), MOD (remainder of integer division), $<_{int}$, $\leq_{int}$, $>_{int}$, $\geq_{int}$ (relational operations)

  comments No lower and upper bound is given for integer numbers since the representation is left open. The usual arithmetic and relational operations are provided. In addition, particular integers can be associated with identifiers (named integers).

  examples Figure 36 (2);

- REAL

values all floating point numbers

operations $+_{real}$ (addition), $-_{real}$ (subtraction), $*_{real}$ (multiplication), / (division), $<_{real}$, $\leq_{real}$, $>_{real}$, $\geq_{real}$ (relational operations)

comments No lower and upper bound is given for floating point numbers since the representation is left open. The usual arithmetic and relational operations are provided.

examples Figure 36 (3);

- BIT STRING

values sequences of bits

operations && (bitwise-AND), || (bitwise-OR), ! (bitwise-NOT), $CONCAT_{bit}$ (concatenation), $SUBSTR_{bit}$ (substring), $LENGTH_{bit}$ (length)

comments The length of the string can be restricted by using the SIZE construct in the type definition. The concatenation operation takes two string arguments and delivers the concatenated string as the result. The substring operation takes one string argument, the starting position, and the length of the substring to be delivered as the result. The length function takes one string argument and delivers the number of bits in the string. In addition, particular bits can be associated with identifiers (named bits).

examples Figure 36 (4);

- OCTET STRING

values sequences of octets

operations $CONCAT_{oct}$ (concatenation), $SUBSTR_{oct}$ (substring), $LENGTH_{oct}$ (length)

comments The length of the string can be restricted by using the SIZE construct in the type definition. The concatenation operation takes two string arguments and delivers the concatenated string as the result. The substring operation takes

```
(1)  D_bit_type    ::= BOOLEAN

         d1 D_bit_type ::= TRUE
         d2 D_bit_type ::= FALSE

         ¬ d1    = FALSE
         d1 ∨ d2 = TRUE
         d1 ∧ d2 = FALSE

(2)  Sequence_numbers ::= INTEGER
     Window_sizes     ::= INTEGER {small_default(2), large_default(8)}

         p_s Sequence_numbers ::= 1
         p_r Sequence_numbers ::= 0
         w_s Window_sizes     ::= small_default

         p_s + p_r              = 1
         ( (p_s - p_r) < w_s ) = TRUE
         ( (p_r + 1) MOD w_s ) = 1

(3)  Throughputs ::= REAL

         tp1 Throughputs ::= (48, 10, 2)
         tp2 Throughputs ::= (12, 10, 2)

         tp1 * (2, 10, 0) = (96, 10, 2)
         tp1 / tp2        = ( 4, 10, 0)

(4)  Packet_type_identifier ::= BIT STRING (SIZE(8))
                                {bit7(7), bit6(6), bit5(5), bit4(4),
                                 bit3(3), bit2(2), bit1(1), bit0(0)}

         pti Packet_type_identifier ::= '00001011'B

         LENGTH(pti)  = 8
         ! pti        = '11110100'B
         pti || ! pti = '11111111'B

(5)  User_data ::= OCTET STRING (SIZE(0..128))

         ud1 User_data ::= '68656C6C6F'H
         ud2 User_data ::= '2021'H

         LENGTH(ud1)        = 5
         LENGTH(ud2)        = 2
         SUBSTR(ud1, 1, 2) = '6865'H
         CONCAT(ud1, ud2)   = '68656C6C6F2021'H

(6)  P_state_type ::= {p1(1), p2(2), p3(3), p4(4), p5(5), p6(6), p7(7)}

         p_state P_state_type ::= p1
```

Figure 36. Simple type examples

	one string argument, the starting position, and the length of the substring to be delivered as the result. The length function takes one string argument and delivers the number of octets in the string.
examples	Figure 36 (5);

- ENUMERATED

values	as given in the enumeration list
operations	-
comments	Each value of an enumerated type is identified by a name and associated with an integer number (encoding information).
examples	Figure 36 (6);

- NULL

values	null
operations	-
comments	The 'empty' type in ASN.1. The value 'null' indicates no contents. It is used as a placeholder for fields that have to be refined by other type definitions.
examples	-

These simple types are similar to most simple types in specification and programming languages. The indices of operation names are omitted in specifications (overloading) because the operations can always be identified by the context in which they are used. A number of character string types (PrintableString, IA5String etc.) are derived from the general OCTET STRING type and predefined in [ISO8824]. The same operations as defined for OCTET STRING are also defined for these character string types.

Structured types

Structured types can be defined in X-ASN.1 by using *type constructors* that facilitate the construction of data structures frequently required in communication systems. Complex types can be formed by combining existing types in various ways.

The following type constructors are provided by X-ASN.1:

- SEQUENCE

values Given a number of types, the constructed type contains all values being tuples of values with one component from each given type (*cartesian product*).

operations $\bullet_{seq,sel}$ (selection), IS_PRESENT_{seq} (for optional fields)

comments Types constructed this way are similar to *records* in programming languages. The order of the components is important, but components may be declared as being optional by using the OPTIONAL construct (optional components may be present or absent in values of such types). Particular components can be accessed with the selection operation by using the individual selector identifiers given in the type definition (dot notation). Since selector identifiers may be omitted in X-ASN.1 type definitions, the position of the component in the type definition can be used as an alternative selector (starting with position number 1 for the first component). Given a particular selector as an argument, the operation IS_PRESENT can be used to ask for the presence of the corresponding component for a given value. Selection of an absent field yields an undefined result.

examples see Figure 37;

- SET

values Given a number of types, the constructed type contains all values being sets of values with one element from each given type.

```
CALL_REQUEST ::= [11] IMPLICIT SEQUENCE {
      header                      Packet_header,
      packet_type_identifier      BIT STRING (SIZE(8)) DEFAULT '00001011'B,
      calling_dte_address_length  INTEGER,
      called_dte_address_length   INTEGER,
      called_dte_address          BCD_STRING (SIZE(0..15))    OPTIONAL,
      calling_dte_address         BCD_STRING (SIZE(0..15))    OPTIONAL,
      facility_length             INTEGER,
      facilities                  FACILITIES                  OPTIONAL,
      user_data                   OCTET STRING (SIZE(0..128)) OPTIONAL
   }

     cr_pdu CALL_REQUEST ::= { header     { q_bit                       FALSE,
                                            d_bit                       FALSE,
                                            modulo_indicator            modulo_8,
                                            logical_channel_group_number 0,
                                            logical_channel_number      1
                                          },
                               packet_type_identifier      '00001011'B,
                               calling_dte_address_length  5,
                               called_dte_address_length   5,
                               called_dte_address          "12345",
                               calling_dte_address         "67890",
                               facility_length             0,
                               user_data                   '68656C6C6F'H
                             }

     cr_pdu.packet_type_identifier        = '00001011'B
     cr_pdu.header.logical_channel_number = 1
     IS_PRESENT(cr_pdu, facilities)       = FALSE


DATA ::= [0] IMPLICIT SEQUENCE {
      header                      Packet_header,
      p_r                         INTEGER,
      more_bit                    BOOLEAN,
      p_s                         INTEGER,
      packet_type_identifier      BIT STRING (SIZE(1)) DEFAULT '0'B,
      user_data                   OCTET STRING (SIZE(0..4096))
   }

Packet_header ::= SEQUENCE {
      q_bit                       BOOLEAN,
      d_bit                       BOOLEAN,
      modulo_indicator            ENUMERATED { modulo_8 (1), modulo_128 (2) },
      logical_channel_group_number INTEGER,
      logical_channel_number      INTEGER
   }
```

Figure 37. Type examples constructed with SEQUENCE

operations $\bullet_{set,\,sel}$ (selection), IS_PRESENT$_{sel}$ (for optional fields)

comments Types constructed this way are similar to *records* in programming languages, but the order of the components is not important. Components may be declared as being optional by using the OPTIONAL construct. Particular components can be accessed with the selection operation by using the individual selector identifiers given in the type definition (dot notation). Since selector identifiers may be omitted in X-ASN.1 type definitions, the position of the component in the type definition can be used as an alternative selector (starting with position number 1 for the first component). Given a particular selector as an argument, the operation IS_PRESENT can be used to ask for the presence of the corresponding component for a given value. Selection of an absent field yields an undefined result.

examples see Figure 38;

```
N_CONNECT_request := SET {
      called_address                          OCTET STRING,
      calling_address                         OCTET STRING,
      receipt_confirmation_selection          BOOLEAN,
      expedited_data_selection                BOOLEAN,
      quality_of_service                      Network_Quality_of_Service,
      user_data                               OCTET STRING
   }

     n_cr_sp N_CONNECT_request ::= { called_address                 '63616C6C6564'H,
                                     calling_address                '63616C6C696E67'H,
                                     quality_of_service             { ... },
                                     user_data                      '68656C6C6F'H,
                                     receipt_confirmation_selection FALSE,
                                     expedited_data_selection       TRUE
                                   }

     n_cr_sp.receipt_confirmation_selection = FALSE
     n_cr_sp.(4)                            = TRUE

N_DATA_request := SET {
      user_data                               OCTET STRING,
      receipt_confirmation_selection          BOOLEAN
   }
```

Figure 38. Type examples constructed with SET

- <u>SEQUENCE OF</u>

values Given a single type, the constructed type contains all values being sequences of values from this type.

operations $[]_{seq, Ind}$ (indexing), $LENGTH_{seq}$

comments Types constructed this way are similar to dynamic *arrays* in programming languages. Particular elements are accessed by indices (starting with index number 1 for the first element). The LENGTH operation delivers the length of the sequence. An index that is greater than the length of the sequence yields an undefined result.

examples see Figure 39;

- <u>SET OF</u>

values Given a single type, the constructed type contains all values being multisets of values from this type.

```
Data_packets := SEQUENCE OF DATA

    dps Data_packets ::= { { header                 { ... },
                             p_r                    0,
                             more_bit               TRUE,
                             p_s                    1,
                             packet_type_identifier '0'B,
                             user_data              '68656C'H
                           },
                           { header                 { ... },
                             p_r                    0,
                             more_bit               FALSE,
                             p_s                    2,
                             packet_type_identifier '0'B,
                             user_data              '6C6F'H
                           }
                         }

    LENGTH(dps) = 2
    dps[1]      = { header                 { ... },
                    p_r                    0,
                    more_bit               TRUE,
                    p_s                    1,
                    packet_type_identifier '0'B,
                    user_data              '68656C'H
                  }
```

Figure 39. Type example constructed with SEQUENCE OF

operations $\in$ (membership), $\bigcup$ (union), $\bigcap$ (intersection), $\setminus$ (difference),
$\subseteq$ (subset), $\subset$ (proper subset), $LENGTH_{set}$

comments Types constructed this way are similar to *sets* in programming languages. Since X-ASN.1 does not require the elements to be different, the mathematical notion of *multisets* is even more accurate. The usual set operations are defined on values of such types. The LENGTH operation delivers the number of elements in the set.

examples see Figure 40;

- CHOICE

values Given a number of types, the constructed type contains all values being an alternative from one of the given types (*discriminated union*).

operations $\bullet_{cho,\, sel}$ (selection), $IS\_ALTERNATIVE_{cho}$

comments Types constructed this way are like *variant records* in programming languages. Particular alternatives are accessed with the selection operation by using the individual selector identifiers given in the type definition. Since selector identifiers may be omitted in X-ASN.1 type definitions, the position of the component in the type definition can be used as an alternative selector (starting with position number 1 for the first component). Given a particular selector as an argument, the operation IS_ALTERNATIVE can be used to ask whether the corre-

```
Data_primitives := SET OF N_DATA_request

    dsps Data_primitives ::= { { user_data '68656C6C6F'H,
                                 receipt_confirmation_selection TRUE } }

    LENGTH(dsps) = 1
    { user_data '68656C6C6F'H, receipt_confirmation_selection TRUE } ∈ dsps = TRUE
```

Figure 40. Type example constructed with SET OF

sponding alternative was chosen for a given value. Accessing other alternatives yields an undefined result.

examples see Figure 41.

The X-ASN.1 type constructors allow the flexible definition of the abstract structure of data objects being exchanged between entities in a communication system. The constructs provided cater for the fact that some parameters of SPs and fields of PDUs are often optional, vary in size, may come in a different order, or may take several alternative forms. The operations defined in combination with the type constructors are generic and predefined for all types built with these constructors. The indices of operation names are omitted in specifications (overloading). The notion of *parameterized types* (or *polymorphic types*) is also often used for such a typing mechanism [Card85].

A concrete representation for values of X-ASN.1 defined types is given separately (supporting data abstraction) by a set of encoding rules (see section 3.1.3) and no specific implementation of the operations is required. The type definitions in X-ASN.1 may be nested (components of structured types may be structured again) and recursive types can be defined as well (optional components of structured types can be of the same type again). A number of useful structured types is predefined by ASN.1 (GeneralizedTime, ObjectDescriptor etc.) and provided to users of the notation. For details, see [ISO8824].

Subtypes

Subtypes comprise a subset of the value set of a parent type (called *base type* in the following). Different constructs exist for the specification of those subsets:

- multiple values

  values enumerated following the base type

  examples Figure 42 (1);

```
X25_PDU := CHOICE {
      call_request                CALL_REQUEST                ,
      incoming_call               INCOMING_CALL               ,
      call_accepted               CALL_ACCEPTED               ,
      call_connected              CALL_CONNECTED              ,
      clear_request               CLEAR_REQUEST               ,
      clear_indication            CLEAR_INDICATION            ,
      clear_confirmation          CLEAR_CONFIRMATION          ,
      data                        DATA                        ,
      interrupt                   INTERRUPT                   ,
      interrupt_confirmation      INTERRUPT_CONFIRMATION      ,
      receive_ready               RECEIVE_READY               ,
      receive_not_ready           RECEIVE_NOT_READY           ,
      reset_request               RESET_REQUEST               ,
      reset_indication            RESET_INDICATION            ,
      reset_confirmation          RESET_CONFIRMATION          ,
      restart_request             RESTART_REQUEST             ,
      restart_indication          RESTART_INDICATION          ,
      restart_confirmation        RESTART_CONFIRMATION        ,
      diagnostic                  DIAGNOSTIC                  ,
      reject                      REJECT                      ,
      registration_request        REGISTRATION_REQUEST        ,
      registration_confirmation REGISTRATION_CONFIRMATION }


     x25_pdu X25_PDU ::= cr_pdu
       --see Figure 37

     x25_pdu.cr_pdu.packet_type_identifier ='00001011'B
     IS_ALTERNATIVE(x25_pdu, call_request) = TRUE


NETWORK_SPS := CHOICE {
      n_connect_request               N_CONNECT_request               ,
      n_connect_indication            N_CONNECT_indication            ,
      n_connect_response              N_CONNECT_response              ,
      n_connect_confirm               N_CONNECT_confirm               ,
      n_data_request                  N_DATA_request                  ,
      n_data_indication               N_DATA_indication               ,
      n_data_acknowledge_request      N_DATA_ACKNOWLEDGE_request      ,
      n_data_acknowledge_indication N_DATA_ACKNOWLEDGE_indication ,
      n_expedited_data_request        N_EXPEDITED_DATA_request        ,
      n_expedited_data_indication     N_EXPEDITED_DATA_indication     ,
      n_reset_request                 N_RESET_request                 ,
      n_reset_indication              N_RESET_indication              ,
      n_reset_response                N_RESET_response                ,
      n_reset_confirm                 N_RESET_confirm                 ,
      n_disconnect_request            N_DISCONNECT_request            ,
      n_disconnect_indication         N_DISCONNECT_indication         }
```

Figure 41. Type examples constructed with CHOICE

- INCLUDES

  values including all values from the base type

  examples Figure 42 (2);

- value range

  values all values from the base type in the specified range

  examples Figure 42 (3);

- SIZE

  values all values from the base type satisfying the size requirements

  examples Figure 42 (4);

- FROM

  values enumerated following the base type (which must be a string)

  examples Figure 42 (5);

- WITH COMPONENTS

  values as given for the base type with or without optional components

  examples Figure 42 (6).

```
(1)  Default_window_sizes := Window_sizes(2 | 8)

(2)  Non_X25_window_sizes := INTEGER(INCLUDES Default_window_sizes | 16 | 64)

(3)  Modulo_8_sequence_numbers   := 0..7
     Modulo_128_sequence_numbers := 0..127

(4)  User_data ::= OCTET STRING (SIZE(0..128))

(5)  Numbers := IA5String
                (FROM("0" | "1" | "2" | "3" | "4" | "5" | "6" | "7" | "8" | "9"))

(6)  INCOMING_CALL := CALL_REQUEST WITH COMPONENTS {calling_dte_address PRESENT}
```

Figure 42. Subtype examples

The operations on subtypes are defined by restricting the operations defined on the base type to the subset of values given by the subtype. Specifying the set of values for an SP parameter or PDU field as a subtype allows more rigorous static type checking to be done.

3.1.2 Type Equivalence

For the definition of equivalence between two X-ASN.1 defined types a weak form of *name equivalence* is used. Pure name equivalence means that two types are equivalent if and only if they have the same name. This is opposed to *structural equivalence*, where two types are equivalent if they have the same underlying structure [Clea86].

For X-ASN.1 the principle of name equivalence has been weakened as follows:

- types are also equivalent if they represent just renamings of the same type;
- subtypes are also equivalent to their base type or to other subtypes if they share the same base type;

With this definition of type equivalence, two given values of equivalent types may be tested for equality. The operation = **(equality)** is defined for all X-ASN.1 defined types and returns a boolean result.

The benefits of ASN.1 as a type definition language for FDTs to be used in the area of communication systems have been recognized by other authors as well (for example [Boch90a] [Thom90]). In [Boch90a], a detailed description of the principles employed in a tool performing a translation from ASN.1 to ACT ONE is given. A formal definition of denotational semantics for ASN.1 with operations is provided in [Thom90]. The functions described give meaning to ASN.1 types (and associated operations) by mapping them to algebraic type definitions in ACT ONE. This work has not been duplicated in this thesis.

3.1.3 Encoding Rules

For each X-ASN.1 type information is given either explicitly or implicitly on how to encode values of that type. This is achieved by using tags and constitutes an important part of the transfer syntax. A tag consists of two parts, a **tag class** and a **tag number**. Tag numbers may be any positive integer number, whereas tag classes always belong to one of the following classes:

UNIVERSAL Universal tags are the implicit tags defined by the ASN.1 standard for each simple type and for each type constructor (for example BOOLEAN has the tag "UNIVERSAL 1" and types constructed with SEQUENCE have the tag "UNIVERSAL 16"). Universal means that tags of this kind are unique in all contexts.

APPLICATION Application tags are defined by a particular application protocol standard. The tags chosen for the defined types are unique in the scope of that application, but may be reused in other applications.

PRIVATE Tags belonging to this class have been assigned by private parties and are unique in some private domain only.

context specific Tags belonging to this class are specific to the context they appear in, for example, in a type definition or a module. These tags are denoted by the tag number only, the tag class being omitted.

In the explicit form, tags are written in square brackets and attached to types in the type definition. If IMPLICIT is used in the type definition only the explicit tag is encoded.

For the mapping between an abstract syntax specified with ASN.1 and a concrete transfer syntax, encoding rules are used. Different sets of encoding rules are possible, leading to different concrete representations of an abstract syntax. Currently, ISO has standardized one set of encoding rules, the so-called *basic encoding rules (BER)* for ASN.1 [ISO8825]. In order to illustrate the way encoding rules work, the BER are explained briefly.

In BER, the structure of encoded values can take two forms: *definite length* or *indefinite length*. Both forms are depicted in Figure 43. Each form can be separated into three parts: (1) type identification, (2) length information, (3) content. In definite length encoding, the actual length of the contents field is calculated and precedes the contents field. If the length is smaller than 127, one byte is used to encode the length, otherwise several bytes are used (the first byte indicates that the length is encoded in the following bytes). If definite length encoding is not appropriate (for example, for strings or structured values), the indefinite length form is chosen. In indefinite length encoding, the length field preceding the contents field indicates that the contents field is terminated by a special "end-of-contents" pattern. The actual length of the contents field is not calculated in advance.

The type identification part for both encoding forms is structured as shown in Figure 44. The first two bits of the identifier octet are used to encode the tag class ("00" for UNIVERSAL, "01" for APPLICATION, "11" for PRIVATE and "10" for context-specific). The next bit indicates whether *primitive encoding* or *constructed encoding* is used. In primitive encoding, the contents field represents an encoded value of a simple type according to encoding rules for all simple types. In constructed encoding, the contents field represents a

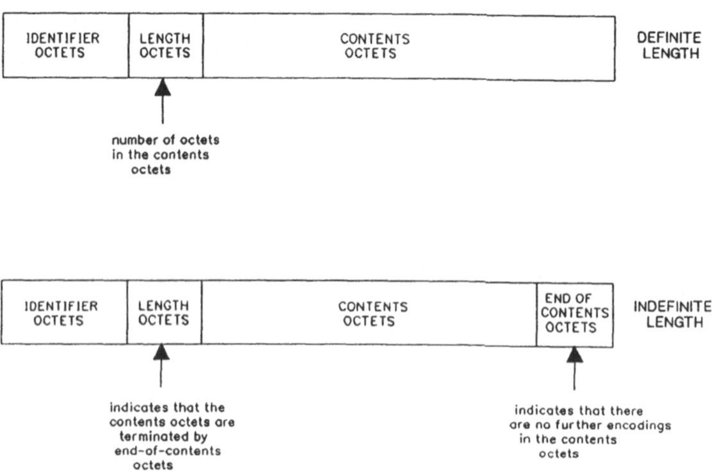

Figure 43. Basic encoding rules

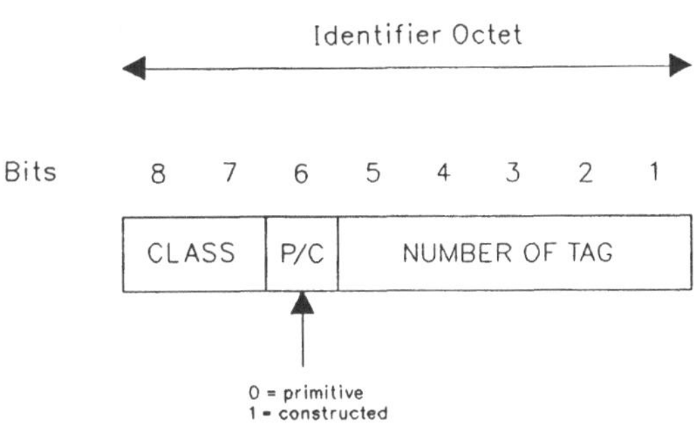

Figure 44. Encoding of the identifier field

value of a constructed type and is again of the form "identification-length-content" (the encoding rules are applied recursively). Finally, the last five bits of the identifier octet contain the tag number distinguishing types of the same tag class. If the tag number does not fit into five bits, an extended form of encoding is used.

Thus, encoding rules define a mapping between abstract structures and concrete representations (in both directions). The BER are completely based on the simple types and on the type constructors of the ASN.1 notation. The structure of the data objects described in the abstract form guides the encoding without using any semantic information about the contents of the fields. Encoding/decoding routines may be implemented in a generic fashion or may be generated individually for particular ASN.1 specifications to automate the implementation step. On the other hand, there may be protocols using a very specific encoding of PDUs. In the X.25 PLP, for example, tags have to be put into the third byte of an encoded packet for packet type identification and into the first byte of each encoded facility for facility identification [Sche90]. Some integers are encoded in 8 bits and 4 bits, for others only 2 bits have to be used. Binary coded digits require 4 bits and all boolean values are encoded in one bit. Such encoding rules are difficult to reuse, but abstraction from the representation is nevertheless achieved and facilitates the specification task from a functional point of view.

3.2 Communicating Rule Systems (CRS)

A communication system composed of several protocol entities in a number of layers is represented in a CRS specification by several hierarchically ordered **rule systems**. These rule systems run in parallel and establish events at **gates** to achieve synchronization. The events occurring at gates may convey parameters for the purpose of communication between rule systems. Gates are abstract means to symbolize the location where the events occur and they are the only objects that can be shared between rule systems. The behavior of a rule system is defined by the possible orderings of externally visible events at gates.

In order to describe the externally visible behavior of a rule system, the extended finite state machine (EFSM) model is used. Corresponding to that model, a rule system consists of a set of state **objects** and a set of **rules** defining state transitions. Each rule is specified by two first order predicate logic (FOPL) formulas characterizing the *start condition* and the *effects* of the rule. In start condition predicates, communication events may be requested at gates, whereas in effects predicates, communication events may be offered.

An example of a communicating rule system configuration is depicted in Figure 45. These rule systems model two network service users, two network entities, and the network. Each rule system consists of a set of objects and a set of rules (depicted as cyclic arrows). The network service is provided to the rule systems Network_user1 and Network_user2 at the gates NSAP1 and NSAP2 by the combination of the rule systems Network_entity1, Network, and Network_entity2. The gates are used to model network SAPs and internal interaction points (Packets1, Packets2). The configuration of rule systems and gates corresponds to the picture used in [ISO8878] (see Figure 9 in chapter 2) for illustrating the use of the X.25 PLP to provide the OSI connection-oriented network service.

The rule systems Network_entity1 and Network_entity2 represent two instances of an X.25 DTE. Thus, the behavior of each of the rule systems Network_entity1 and Network_entity2 corresponds to the specification example used throughout chapter 2. A CRS specification of the configura-

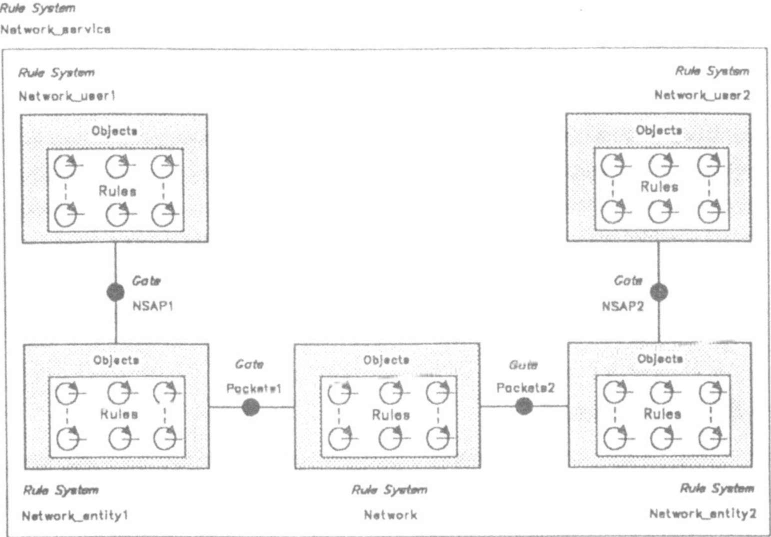

Figure 45. CRS network service configuration

tion depicted in Figure 45 is shown in Figure 46. The rule system `Network_service` is composed of several other rule systems and gates that are instances of parameterized rule system and gate specifications. Several rule systems may share a gate if the same gate instance is given to the rule system instances as an actual parameter. By composing rule systems with common gates, a large system can be described as a hierarchy of interacting smaller components. In this way, ***compositionality*** is supported

```
RULE_SYSTEM Network_service;

   STATE

   NSAP1, NSAP2        : Network_service_access_point       ;
   Packets1, Packets2  : X_25_network_access_point          ;
   Network_user1       : Network_service_user (NSAP1)       ;
   Network_user2       : Network_service_user (NSAP2)       ;
   Network_entity1     : X_25_DTE (NSAP1, Packets1, 1    )  ;
   Network_entity2     : X_25_DTE (NSAP2, Packets2, 4095)   ;
   Network             : X_25_network (Packets1, Packets2)  ;

END_RULE_SYSTEM Network_service;
```

Figure 46. CRS specification of the network service configuration

at the rule system level. The behavior of the whole system is determined by the behavior of the components.

In order to describe the behavior of the active components, several smaller rule systems and gates may be composed again. When the decomposition is completed, the behavior of a rule system will have been specified using only rules and passive state objects. The object types and the various constructs for specifying rule systems, rules, and gates that are available in CRS are explained in more detail in the following. The same example as in chapter 2 is used for illustration by refining the rule system specification X_25_DTE.

3.2.1 Rule Systems

Rule systems are active parallel components in a CRS specification. The behavior of a rule system is defined in a rule system specification as outlined in Figure 47 (using an extended Backus-Naur Form). A rule system specification can be parameterized and it can be used for creating rule system instances with different actual parameters substituted for the formal parameters in the formal parameter list.

Each rule system instance is based on an EFSM. Major and minor state objects are declared following the keyword STATE and may be of different types. The possible transitions of the EFSM are specified using rules that are defined in the RULES-part of a rule system specification. In order to express the start condition and the effects of a rule, auxiliary objects and predicates may be used that are listed following the keywords DECLARATIONS and DEFINITIONS. Useful constant values can be defined following the keyword CONSTANTS. A number of operators are provided by CRS that allow the specification of relations between rules acting independently on the rule system state. These relations are defined in the SYN-part of a rule system specification.

In Figure 48, an outline of a rule system specification is given that models a network entity (see Figure 46) as an X.25 DTE. The gates representing the network SAP to which the entity is attached and the internal interaction point to the network are formal parameters (Nsap, Packets). An additional formal parameter is used for associating instances with a particular logical

```
RULE_SYSTEM <rule_system_name> (<formal_parameters>);

 CONSTANTS
             {<constant_name>              = <constant_value> ;}
 STATE
             {<state_variable_name>        : <state_variable_type> ;}
 DECLARATIONS
             {<declaration_variable_name> : <declaration_variable_type> ;}
 DEFINITIONS
             {<auxiliary_predicate_name> (<formal_parameters>) ⇔ <formula> ;}
 RULES
               ...
 SYN
             {<rule_name> <synchronization_operator> <rule_name> ;}

END_RULE_SYSTEM <rule_system_name>;
```

Figure 47. Syntax of rule system specifications

```
RULE_SYSTEM X_25_DTE ( Nsap    : Network_service_access_point;
                       Packets : X_25_network_access_point;
                       LC      : Logical_channel_number );

   STATE

      r_state           : R_state_type INITIALLY r1;
      p_state           : P_state_type INITIALLY p1;
      d_state           : D_state_type INITIALLY d1;

      next_to_send      : Sequence_numbers;
      last_ack_received : Sequence_numbers;
      next_to_receive   : Sequence_numbers;
      last_ack_sent     : Sequence_numbers;

      send_queue        : DataQueue;
      receive_queue     : DataQueue;
      ...

   DECLARATIONS

      N_CR_SP         : N_CONNECT_request;
      CR_PDU          : CALL_REQUEST;
      ...

   RULES
      ...

END_RULE_SYSTEM X_25_DTE;
```

Figure 48. CRS specification of X.25 DTE

channel (LC). The state objects r_state, p_state, d_state determine the major states of the rule system corresponding to the "restart", "call setup and clearing", and "data transfer" states used in [ISO8208]. Minor state objects are declared for holding flow control information, user data to be sent and received etc. Auxiliary objects are required for event parameters, intermediate results etc. Some examples are given in Figure 48.

The object types and the rule paradigm employed in CRS are further detailed in the following two sections.

Objects

In CRS, objects may be used as state objects or as local objects of rules[6]. The scope of state objects extends over the whole rule system, whereas the scope of local objects is restricted to the start condition and effects formula of a particular rule. Each object in CRS is of an object type. Depending on the nature of the object, object types describe the abstract structure, the possible operations, or the observable behavior of the object. According to their object types, objects used in CRS specifications can be classified into

1. **passive objects**, being
   - X-ASN.1 defined types or
   - abstract data types (ADTs);
2. **active objects**, being
   - gate instances or
   - rule system instances.

Rule systems using only passive state objects are at the lowest level of a rule system hierarchy defined by a CRS specification and are called **basic rule systems**. Rule systems that are composed of active objects are called **communicating rule systems**. This technical distinction is important for defining the semantics of CRS in section 3.3.

[6] The term 'object' is used in a broader scope here, and not restricted to the object-oriented interpretation.

```
<rule_name>;
  COND:       <start_condition_formula> ;
  EFFECTS:    <effects_formula>         ;
```

Figure 49. Syntax of rules

The rule systems Network_entity1 and Network_entity2 are examples of basic rule systems. The state objects are of X-ASN.1 types (see section 3.1), except for send_queue and receive_queue, which are instances of the ADT DataQueue (see the rule system specification for X_25_DTE). How ADTs are specified in CRS is explained in section 3.2.4. The rule system Network_service specified in Figure 46 is an example of a communicating rule system.

Rules

The syntax of CRS rules is shown in Figure 49. Each rule is identified by a name and described by two formulas specifying the start condition and the effects of a rule. The start condition is a predicate on the rule system state, event occurrences, and requested event parameters. The effects predicate defines a relation between the rule system state before the effects are established ("old state") and the state afterwards ("new state"). A prime (') is used in the effects formula when referring to an object of the old state. Additionally, the effects formula describes event occurrences, event offers, and offered event parameters. A rule may be started if the start condition holds. A rule that has been started but not yet ended is called "running". When a rule ends, actions establishing the specified effects are performed atomically. Starting and ending a rule are two separate actions, occurring at different points in time. After its termination, a rule may be started again if the start condition is satisfied. As an example, a small part of a rule describing the initiation of a connection establishment by a network service user (resulting in the initiation of the X.25 PLP call setup procedure) is given in Figure 50. In the start condition of the rule n_con_req, the

```
n_con_req;
  COND:       p_state = p1   ∧  Nsap.N_CONNECT_req (N_CR_SP);
  EFFECTS:    p_state = p2   ∧  Packets.Call_request (CR_PDU)
                             ∧  CR_PDU.user_data = N_CR_SP.user_data;
```

Figure 50. CRS specification of X.25 PLP call setup initiation

N_CONNECT_req event is requested at the gate Nsap (see Figure 52 for the syntax of event expressions) and requirements on the object p_state are expressed. In the effects, the Call_request event is offered at the gate Packets and the values of event parameters are specified; a new value for p_state is given. In this example, the local objects N_CR_SP and CR_PDU have been used as event parameters. The scope of these objects is local to the rule n_con_req, but the value of N_CR_SP is remembered from the start of the rule until execution of the actions establishing the effects. Thus, the values of local objects are available for specifying the effects of a rule and for describing offered parameter values, depending on received event parameter values and state objects when the rule was started. Components of objects of structured X-ASN.1 types are accessed using the X-ASN.1 selection operations as defined in section 3.1.

In order to describe how the X.25 PLP is used for establishing a connection at the request of a network service user, the rule n_con_req has to be specified in much more detail and several other rules are required for processing the responses from the remote side (see [Mang90]). Rules represent autonomous pieces of knowledge about a rule systems' behavior. In the area of communication systems, rules are considered to be an excellent means for capturing protocol knowledge and for supporting ***intelligibility*** of large protocol specifications. Each rule can be understood on its own and does not immediately affect other rules, which reduces the complexity of the specification task and simplifies extended or modified specifications. If several rules can be applied in a certain state, the choice of which one to select is nondeterministic, which realizes ***control abstraction***. By using predicate logic, start conditions and effects of rules are specified in an implementation-independent way.

Normally, most of the internal actions of basic rule systems are sequential in nature. Therefore, rules in a basic rule system are mutually exclusive by default. However, internal parallelism of rules inside one rule system is sometimes useful. In CRS, three operators are provided for specifying internal synchronization requirements. These operators can be used to explicitly allow for parallel or privileged execution of rules. The

- **semiparallel operator** (|-) is used for specifying that a rule can be started even if another rule is already running;
- **parallel operator** (||) is used for specifying that either rule can be started even if the other one is already running ("semiparallelism in both directions");
- **prior operator** ($\prec$) is used for specifying that one rule has a higher priority if the start conditions of both rules are satisfied.

An example of the application of the internal synchronization operators is given in Figure 51. In the formal description of the X.25 PLP, rules describing the process of fragmenting, sending, receiving, and reassembling data are allowed to run in parallel (duplex data transfer). Whenever it is possible to send or receive normal data and expedited data, the rules for expedited data transfer are given priority.

Internal synchronization in CRS is not only useful for specifying protocols, but for distributed systems in general because rule systems are regarded as distribution units. Whereas rule systems use gates to communicate, rules inside a rule system are able to exchange information by using state objects. The first mechanism may be used to model message passing, whereas the second describes a form of shared memory. The distinction between these two communication mechanisms is crucial in distributed system design (see for example [Nehm87]).

3.2.2 Gates

In CRS, gates are special objects shared by rule systems for the purpose of communication. They may be considered as an abstract notion for the

```
SYN
fragment_data           ||      send_data;
send_data               ||      receive_data;
receive_data            ||      reassemble_data;
send_expedited_data     ≺       send_data;
receive_expedited_data  ≺       receive_data;
```

Figure 51. CRS specification of X.25 PLP duplex and expedited data transfer

```
<gate_name> . <event_name> ( <parameters> ) .
```

Figure 52. Syntax of event expressions

location where communication events occur[7]. More than two rule systems may share one gate, but events always occur between two rule systems. In order to refer to an event in a specification an *event expression* is used (see Figure 52). An event expression in the start condition of a rule represents an event request, whereas an event expression in the effects part represents an event offer. An event of a certain type occurs at a certain gate if an event offer of that type is made by one rule system and an event request of that type is made by another rule system. Additionally, the offered event parameters have to satisfy the constraints imposed by the requesting rule system. If the event occurs, the parameter values are passed to the requesting rule system.

In a gate specification, possible event types and their parameter types are defined, employing the syntax shown in Figure 53. In the **DECLARATIONS**-part, the data types of event parameters are given. In the **EVENTS**-part, possible event types are listed together with their individual event parameters. Several gate instances may be created and allow for the occurrence of events of the event types defined in the gate specification. As examples, gate specifications for the gate types

```
GATE (A)SYNCHRONOUS <gate_name> ;

 DECLARATIONS
            {<parameter_name> : <parameter_type_name> ;}

 EVENTS
            {<event_name> ( <event_parameters> ) ;}

END_GATE <gate_name> ;
```

Figure 53. Syntax of gate specifications

---

7 Other FDTs also use the term 'gate' (LOTOS), or similar terms like 'label' (CCS), 'channel' (Estelle) or 'port' (TCSP).

```
GATE SYNCHRONOUS
   Network_service_access_point;

   DECLARATIONS
   N_CR_SP : N_CONNECT_request;
   N_CI_SP : N_CONNECT_indication;
   ...

   EVENTS
   N_CONNECT_req (N_CR_SP);
   N_CONNECT_ind (N_CI_SP);
   ...

END_GATE Network_service_access_point;
```

```
GATE SYNCHRONOUS
   X_25_network_access_point;

   DECLARATIONS
   CR_PDU : CALL_REQUEST;
   IC_PDU : INCOMING_CALL;
   ...

   EVENTS
   Call_request  (CR_PDU);
   Incoming_call (IC_PDU);
   ...

END_GATE X_25_network_access_point;
```

Figure 54. CRS specification of X.25 PLP gates

Network_service_access_point and X_25_network_access_point are given in Figure 54. The data types of the event parameters are specified in X-ASN.1, as illustrated in section 3.1.

In an extensional specification, possible events at gates are an important piece of information. Gate specifications define the ***structure*** of the interaction domain of rule systems sharing gate instances. In the area of communication services and protocols, gates are used to model SAPs or internal interaction points of protocol entities.

Rule systems may communicate at gates either synchronously or asynchronously. The communication mode is defined in the gate specification (see Figure 53). At synchronous gates, only event occurrences are seen by an external observer. When the event offer and the event request coincide, the offering and requesting rule systems participate in a two-way rendezvous. Asynchronous communication mode is characterized by buffering events at gates. In this case, the event offer and the event occurrence can be observed. Event offers are buffered as sets, assuming infinite buffering capacity.

Providing both synchronous and asynchronous communication increases the ***expressive power*** of an FDT (interpreting this criterion as explained in chapter 2). In a technique providing only the synchronous communication mode, asynchronous communication between components can be modelled by introducing an explicit buffering component. However, it contributes to ***abstraction*** and ***intelligibility*** of specifications if communication primitives for both communication modes are provided, so that the details of such buffering components can be omitted.

3.2.3 Structuring Mechanisms

The rule system concept is the preferred structuring mechanism in the first phase of designing a complex system and it may be repeatedly applied to identify loosely coupled system components [Velt90]. At some stage of the design, however, basic rule systems remain and behavior is specified by using rules. These rules can be given in the `RULES`-part of a rule system specification as a flat list (*flat rule system*).

The behavior of a basic rule system may still be complex. The experience gained during specification case studies for existing protocols revealed a lack of structuring mechanisms that can be applied to the set of rules in a rule system specification [Monn89] [Mang90] [Velt90]. As a result, additional concepts have been developed for structuring the rule set. In particular, the following mechanisms are provided by CRS:

- **definitions** (auxiliary predicates),
- **contexts,**
- **views.**

If these mechanisms are applied to specifications, *structured rule systems* are obtained. Structured rule system specifications are easier to develop and more intelligible than flat ones. The semantics of structured rule systems is defined by describing how an equivalent flat rule system can be constructed that results from expanding all definitions and resolving all contexts and views. These mechanisms are explained in the following.

Definitions

Auxiliary predicates are defined in the `DEFINITIONS`-part of a rule system specification. A predicate name is defined and a list of formal parameters may be given. The body of an auxiliary predicate definition is specified as a formula with state objects and parameters as free variables.

An example is given in Figure 55: an auxiliary predicate is defined that holds if the sliding window used for flow control in the X.25 PLP is open[8]. Attention must be paid to the fact that the sequence numbers marking the upper and the lower edge of the window are increased "modulo 8" or "modulo 128" (as indicated by the state object `cycle_size`). Since a sliding window is maintained for both directions of data transfer in the X.25 PLP, the auxiliary predicate `window_open` can be used to specify necessary conditions for the transmission and reception window.

Auxiliary predicates can be used in start condition and effects formulas of rules and achieve more compact and intelligible specifications. If meaningful names are chosen, the purpose of the predicate can be deduced from the name of the predicate and rules can be read like prose. The semantics of these definitions is as follows: each occurrence of an auxiliary predicate is textually replaced by the predicate body after substituting formal parameters by actual ones. This is called **expansion** of auxiliary predicates.

Contexts and Views

Contexts and views are two orthogonal structuring mechanisms that can be arbitrarily nested. **Contexts** are used to group logically related rules by extracting common parts of their start condition and effects formulas to form

```
DEFINITIONS

 window_open (upper_window_edge, lower_window_edge, window_size)

 ⇔

  IF
    (upper_window_edge >= lower_window_edge)
   THEN
    (upper_window_edge - lower_window_edge) < window_size
   ELSE
    (upper_window_edge + cycle_size - lower_window_edge) < window_size
  FI
```

Figure 55. CRS specification of X.25 window mechanism

---

[8] Note that the formula ' IF a THEN b ELSE c ' is equivalent to ' $a \wedge b \vee \neg a \wedge c$ '.

```
CONTEXT <context_name> ;
   COND:        <context_condition_formula> ;
   EFFECTS:     <context_effects_formula>   ;
   ...

END_CONTEXT <context_name> ;
```

Figure 56. Syntax of contexts

context condition and effects formulas. The syntax used for contexts is given in Figure 56. If the context condition formula is $\wedge$-connected to the start condition formula, and the context effects formula is $\wedge$-connected to the effects formula of all the rules in the context, an equivalent flat rule system is obtained. This is called **resolution** of contexts.

Different **views** are defined on the set of rules to focus on different aspects in a rule system specification. This is done by splitting start condition and effects formulas of rules into several parts and include these parts under the same rule name in different views defined in the specification. In this way, the complexity of rule start condition and effects formulas can be reduced by separating constraints on different state objects, event occurrences, event parameters etc. into different views. The syntax used for views is given in Figure 57. If the various parts of start condition and effects formulas of the same rules from all views are $\wedge$-connected, an equivalent flat rule system is obtained. This is called **resolution** of views.

As an example of the application of contexts and views, the overall structure of the rule system specification for an X_25_DTE is shown in Figure 58. A top-level structure in terms of contexts is chosen and protocol functions are subdivided into those for Call_setup, Data_transfer, Call_clearing, and Restart, corresponding to the different phases of the X.25 PLP. The

```
VIEW <view_name> ;

   ...

END_VIEW <view_name> ;
```

Figure 57. Syntax of views

```
RULES

  CONTEXT Call_setup ...

    VIEW Main_state_and_events ...
    VIEW User_data_handling ...
    VIEW Addresses ...
    VIEW Facility_handling ...
    VIEW Expedited_data_service ...
    VIEW Receipt_confirmation_service ...
    VIEW Quality_of_service ...
    VIEW Timers_and_counters ...

  CONTEXT Data_transfer ...

    CONTEXT Normal_data ...

      VIEW Main_state_and_events ...
      VIEW User_data_handling ...
      VIEW Receipt_confirmation_service ...
      VIEW Flow_control ...
      VIEW Retransmission ...

    CONTEXT Expedited_data ...

    CONTEXT Reset_service ...

  CONTEXT Call_clearing ...

  CONTEXT Restart ...
```

Figure 58. Context and view structure of the X.25 PLP specification in CRS

context Data_transfer is further structured into the contexts Normal_data, Expedited_data, and Reset_service. The contexts Call_setup and Normal_data are split into several views capturing different aspects of the X.25 PLP (for details see [Mang90]).

An example of how several parts of rules can be spread over different views for the case of data transfer in the X.25 PLP is given in Figure 59. A common condition for all rules describing data transfer aspects of the protocol is given by the requirement on the state object p_state, which must have the value p4. Similarly, the value of the state object d_state must be equal to d1 for all rules describing normal data transfer. The focus in the view Main_state_and_events is on the main state objects (r_state, p_state, d_state) and on event occurrences. The rules send_data and receive_data are shown in the example. The data to be sent is passed by the network service user with the event N_DATA_req, fragmented into pieces according to the packet size used, and inserted into the ADT object send_queue (described by the rule fragment_data which is not shown in Figure 59). The rule send_data is allowed to start as soon as the send_queue contains some data fragments, which is expressed by using the ADT operation empty on the ADT instance send_queue. A DATA packet is sent in the effects part of the rule send_data and the contents of the user_data field is retrieved from the send_queue.

The data from incoming DATA packets is inserted into the ADT object receive_queue as specified by rule receive_data and retrieved for reas-

```
CONTEXT Data_transfer;
 COND: pstate = p4;

  CONTEXT Normal_data;
   COND: dstate = d1;

    VIEW Main_state_and_events;

     send_data;
     COND:      ¬ (send_queue.empty);
     EFFECTS:   Packets.Data(DATA_PDU) ∧ DATA_PDU.user_data = send_queue.remove;

     receive_data;
     COND:      Packets.Data(DATA_PDU);
     EFFECTS:   receive_queue.insert(DATA_PDU.user_data);
     ...
    END_VIEW Main_state_and_events;

    VIEW Flow_control;

     send_data;
     COND:      window_open(next_to_send, last_ack_received, trans_window_size);
     EFFECTS:   DATA_PDU.p_s = 'next_to_send
      ∧ next_to_send = ('next_to_send + 1) MOD cycle_size
      ∧ DATA_PDU.p_r = 'next_to_receive ∧ last_ack_sent = 'next_to_receive;

     receive_data;
     COND:      window_open (next_to_receive, last_ack_sent, recep_window_size)
      ∧ DATA_PDU.p_s = next_to_receive ∧ p_r_valid (DATA_PDU.p_r);
     EFFECTS:   last_ack_received = DATA_PDU.p_r
      ∧ next_to_receive = ('next_to_receive + 1) MOD cycle_size;
     ...
    END_VIEW Flow_control; ...
  END_CONTEXT Normal_data; ...
END_CONTEXT Data_transfer;
```

Figure 59. Contexts and views example for X.25 PLP data transfer specified with CRS

sembly by the rule `reassemble_data` (not shown in Figure 59). The flow control aspects are specified in the view `Flow_control`. An additional requirement for starting the rule `send_data` is expressed in this view by using the auxiliary predicate `window_open` defined in Figure 55. The state objects `next_to_send` and `last_ack_received` marking the upper and lower edge of the transmission window are supplied as actual parameters, together with the state object `trans_window_size`. In the effects part of the rule `send_data` in the view `Flow_control`, the values of the `p_r` and `p_s` fields are specified and the window edges are updated. In a similar way, flow control constraints are specified for the rule `receive_data` and windows are rotated in the effects part. The other protocol aspects that are important for the rules `send_data` and `receive_data` are specified in the other views as shown in Figure 58, but have been omitted in the example.

Contexts and views are powerful mechanisms used to ***structure*** rule systems. The ***formal definition*** of these constructs is achieved by mapping structured rule systems to flat ones obtained by expansion and resolution. Both contexts and views support different forms of ***compositionality*** at the rule level [Velt90]. Contexts are most often used for grouping related protocol functions in rule systems. Views are used for separating the specification of different protocol aspects.

3.2.4 Abstract Data Types

Abstract data types (ADTs) are specified in CRS by employing a state machine model similar to rule systems. An ADT specification defines a set of operations that may be invoked on ADT objects. The operations control the modification of a set of state objects and are the only means for manipulating them. The syntax of ADT specifications is given in Figure 60. The different parts of an ADT specification are similar to those of rule system specifications. However, there is an **OPERATIONS**-part instead of a **RULES**-part. ADT operations may not be executed in parallel. Therefore, a **SYN**-part does not exist. Unlike rule systems, ADTs are passive objects. ADT operations have to be invoked by active objects in the specification and may be called in any order. However, only one operation can be performed at a time. By hiding the internal representation of state objects and the way operations perform the modifications of the objects, ***data abstraction*** is achieved.

```
ADT <adt_name> ( <formal_parameters> );

 CONSTANTS
              {<constant_name>              = <constant_value> ;}
 STATE
              {<state_variable_name>        : <state_variable_type> ;}
 DECLARATIONS
              {<declaration_variable_name> : <declaration_variable_type> ;}
 DEFINITIONS
              {<auxiliary_predicate_name> (<formal_parameters>) ⇔ <formula> ;}
 OPERATIONS
                 ...
END_ADT <adt_name>;
```

Figure 60. Syntax of ADT specifications

State objects of ADTs may only be passive objects according to the classification of section 3.2.1. An operation is specified by two formulas describing the *precondition* and the *effects* of the operation. The syntax used for specifying operations is given in Figure 61. The precondition formula is a predicate on the ADT's state objects and operation input parameters. The effects formula relates the new state reached after execution of the operation and the output parameters to the old state and the input parameters. As in effects parts of rules, primes are used to refer to state objects before execution of the operation.

In the example given in Figure 62, state objects for the ADT `DataQueue` are declared holding the `contents` of the queue, the maximum `size`, `front` and `rear` markers. Two other state objects are used to indicate whether the queue is `empty` or `full` . The operations defined for the ADT `DataQueue` allow for initializing the queue (`init`), inserting and removing data (`insert`, `remove`), and querying the status (`empty`, `full`).

```
<operation_name>;
   PRE:        <precondition_formula> ;
   EFFECTS:    <effects_formula>      ;
```

Figure 61. Syntax of abstract operations

```
ADT DataQueue;

 STATE

  contents            : SEQUENCE OF OCTET STRING;
  size, front, rear   : INTEGER;
  empty, full         : BOOLEAN;

 DECLARATIONS

  data                : OCTET STRING;
  new_size            : INTEGER;
  result              : BOOLEAN;

 OPERATIONS

  init(new_size);
   PRE:        TRUE;
   EFFECTS:    size = new_size ∧ front = 1 ∧ rear = 1
             ∧ empty = TRUE ∧ full = FALSE;

  insert(data);
   PRE:        ¬(full);
   EFFECTS:    contents['rear] = data ∧ rear = ('rear + 1) MOD size
             ∧ full = (front = rear) ∧ empty = FALSE;

  remove() -> data;
   PRE:        ¬(empty);
   EFFECTS:    data = contents['front] ∧ front = ('front + 1) MOD size
             ∧ empty = ( rear = front) ∧ full = FALSE;

  empty() -> result;
   PRE:        TRUE;
   EFFECTS:    result = empty;

  full() -> result;
   PRE:        TRUE;
   EFFECTS:    result = full;

END_ADT DataQueue;
```

Figure 62. CRS specification of ADT DataQueue

3.2.5 Assessment

The key goal of CRS is to model a system's externally observable behavior by employing the EFSM model and predicate logic. The combination of these two well-established approaches and the use of a rule paradigm yields a powerful FDT. Additionally, the ***formal definition*** of CRS is facilitated by these concepts. In section 3.3, the semantics of CRS as used in this thesis is defined by using general *labelled transition systems (LTSs).*

Compared to other techniques employing state machine models, the ***expressiveness*** of CRS has been improved. Synchronous and asynchronous communication primitives are provided for the convenience of specifiers. Different forms of parallelism exist at both the rule system level and the rule level. The data typing facilities provided by X-ASN.1 are tailored towards the needs of communication systems, making them extremely useful in practice. The full power of predicate logic is provided for specifying possible behavior of rule systems. However, restrictions on the form of predicates may be appropriate in order to simplify the development of tools.

Abstraction is achieved with CRS in various ways. The rule paradigm and nondeterministic selection of rules realizes control abstraction in CRS specifications. The abstract structure of data objects is separated from the concrete representation by using X-ASN.1 as the data description technique. The CRS mechanisms allow the definition of ADTs and support data abstraction. The objects and rules of a rule system are only used to specify the externally visible behavior of a system.

Several concepts are provided by CRS to support ***compositionality*** at different levels of granularity. Rule systems can be composed in parallel to form a larger system and the behavior of a single rule system may be defined by composing several contexts and views. Contexts and views are composed of a set of rules that work independently on the rule system state. These mechanisms facilitate the decomposition of complex communication services and the detailed description of communicating components.

Imposing a ***structure*** on CRS specifications is facilitated by rule systems, contexts, and views, and also by using auxiliary predicates and gate specifications. Altogether, a comprehensive set of mechanisms is provided that ease the task of formally specifying complex communication services and protocols. A suitable structure increases the ***intelligibility*** of specifications. However, the intelligibility of formal descriptions depends on the basic paradigms of an FDT. Employing rules is considered to be advantageous in the area of communication protocols in order to bridge the gap between natural language text and the formal specification of protocols.

The ***constructivity*** of CRS specifications is crucial to the ideas presented in chapter 5 of this thesis. The possibility of interpreting specifications by automated tools can be exploited to support the whole development life-cycle of communication systems. In order to provide such support, the specification knowledge has to be adequately represented and offered to development tools. This can be achieved with rules in an elegant way.

3.3 Operational Semantics of CRS

Formal semantics can be defined in various ways, the most common of which are the operational, denotational, and axiomatic styles (see for example [Loec87]). Originating from theoretical work on sequential programs, these mathematical approaches have been adapted and extended by many authors to cope with concurrent systems and communication.

In this section, an operational semantics for CRS is defined based on *labelled transition systems (LTSs)* as the general formalism. Only synchronous communication between rule systems is covered, although asynchronous communication between rule systems could easily be included.

3.3.1 Labelled Transition Systems (LTSs)

LTSs are a particular type of *transition system*, first introduced in [Kell76] as an abstract "conceptual model" for various forms of parallelism. LTSs have been frequently used to define operational semantics of specification languages (see for example [ISO8807] [ISO9074]). Due to the general nature of this formalism, LTSs can be used to model quite different concurrent systems and may serve as the semantic basis for a wide range of specification techniques. LTSs are based on two primitive notions, namely *states* and *transitions*. For each system or model for which some form of global state and some form of indivisible action causing state transitions can be identified, a corresponding LTS can be defined [DeNi87].

Definition 3.3.1.a

A **labelled transition system (LTS)** is defined as LTS = (Q, ACT, $\dashrightarrow$), where

Q is a set of **states**,

ACT is a set of **actions**,

$\dashrightarrow$ is a **transition relation**, $\dashrightarrow \subseteq Q \times ACT \times Q$.

For $(q, a, q') \in \dashrightarrow$ we write $q \xrightarrow{a} q'$.

In general, neither Q nor ACT have to be finite. The elements of the set of states Q are often defined as a cartesian product of several components. If the transition relation $\dashrightarrow$ is a (partial) function $\dashrightarrow$: $Q \times ACT \rightarrow Q$, the resulting LTS is deterministic. Otherwise, there may be several transitions with the same action resulting in different subsequent states [Kell76]. The set of actions ACT may contain single events, but also sets of events as elements. Different semantics of concurrency can be defined in this way.

3.3.2 Basic and Communicating Rule Systems

Parallel components in a CRS specification are given by a finite set RS of rule systems. Two forms of rule system instances are distinguished in RS (see section 3.2.1):

1. **basic rule systems**, using only passive state objects;
2. **communicating rule systems**, using only active state objects.

The active state objects are rule system or gate instances and the passive state objects are objects of X-ASN.1 types or ADT objects. Rule system specifications using state objects of both forms are considered as a shorthand notation for two rule systems, running asynchronously in parallel, with the state components being separated into the two corresponding forms.

Considering a CRS specification hierarchy as a tree of rule system instances, communicating rule systems represent interior nodes of this tree and their behavior is solely determined by their successors; the RULES-part is empty. Basic rule systems represent leaf nodes and their behavior is defined by rules.

```
Example:
The rule system Network_entity1 of Figure 46 is an example of a basic rule system.
The rule system Network_service of Figure 46 is an example of a communicating rule system.
```

States

Let SV(rs) denote the set of state variables of a rule system 'rs', which are declared in the rule system specification as

$sv_1 : type(sv_1);$

...

$sv_n : type(sv_n);$

following the keyword STATE. Let

$$SV(rs) := \{ sv_1, ..., sv_n \}$$

$$StatVal(rs) := D_{type(sv_1)} \times ... \times D_{type(sv_n)}$$

where D_γ denotes the set of values of type γ.

Primes ('') are used in CRS to reference state variables holding values of the rule system state before the actions establishing the effects of a rule are performed, whereas unprimed variables denote the state after execution of these actions. Therefore, the following additional set of variables

$$'SV(rs) := \{'sv_1, ..., 'sv_n\}$$

is introduced, where each $'sv_i$ has the same type as sv_i (i = 1...n).

Let LV(rs) denote the set of local variables defined in a rule system 'rs' in the form

lv_1 : type(lv_1);

...

lv_m : type(lv_m);

following the keyword **DECLARATIONS**. Let

$$LV(rs) := \{lv_1, \ldots, lv_m\}$$

$$LocVal(rs) := D_{type(lv_1)} \times \ldots \times D_{type(lv_m)}.$$

Local variables are mainly used to save the values of event parameters and state variables when a rule is started until the effects of the rule are established. This is useful because starting a rule (checking the start condition) and ending a rule (establishing the effects) are considered two separate actions that might be interleaved with the starting or ending of other rules. However, some local variables are only used in the start condition and effects formula, respectively. In this case, these variables represent auxiliary objects that are introduced to simplify logical expressions.

Example:

```
SV(Network_entity1)       = {r_state, p_state, d_state,
                             next_to_send, last_ack_received,
                             next_to_receive, last_ack_sent,
                             send_queue, receive_queue, ...}

StatVal(Network_entity1) =   R_state_type × P_state_type × D_state_type
                           × Sequence_numbers × Sequence_numbers
                           × Sequence_numbers × Sequence_numbers
                           × DataQueue × DataQueue × ...

LV(Network_entity1)       = {N_CR_SP, CR_PDU, ...}

LocVal(Network_entity1)  = N_CONNECT_request × CALL_REQUEST × ...
```

(see the type definitions given in Figure 36, Figure 37, Figure 38, and Figure 62.)

The behavior of a basic rule system 'rs' is defined by the set of rules specified in the RULES-part. Let

RULES(rs)

denote the set of rules of the rule system 'rs'. Each rule takes the form

```
r;
COND:       F_cond(rs, r) ;
EFFECTS:    F_effects(rs, r) ;
```

where $F_{cond\,(rs,r)}$ is a formula with free variables in $SV(rs) \cup LV(rs)$ and $F_{effects\,(rs,r)}$ is a formula with free variables in $SV(rs) \cup {}'SV(rs) \cup LV(rs)$.

```
Example:
RULES(Network_entity1) = {n_con_req,
                          send_data, receive_data,
                          fragment_data, reassemble_data,
                          send_expedited_data, receive_expedited_data, ...}
```

$F_{cond\,(Network\_entity1,\, n\_con\_req)}$ and $F_{effects\,(Network\_entity1,\, n\_con\_req)}$ are given in Figure 50.

The set of running rules in a rule system 'rs' is denoted by

$RunRules(rs) \subseteq RULES(rs).$

After a rule has been started, the values saved in local variables belong to the execution state of this rule. The set of execution states of rules is defined as

$RunRulesLocVal(rs) := \mathbf{P}(RULES(rs) \times LocVal(rs))$

where **P**(s) denotes the powerset of a set s. In a similar way as for state variables, let 'LV(rs) denote the set of local variables holding values before the actions establishing the effects of a rule are performed and LV(rs) the set of local variables of the effects formula.

The state of a basic rule system 'rs' comprises the values of all state variables and the execution state of all running rules. The set of all possible states Q_{rs} is given by the cartesian product

$$Q_{rs} := \text{StatVal}(rs) \times \text{RunRulesLocVal}(rs).$$

A typical state is denoted as $q_{rs} = (\underline{sv}, rrlv)$.

A communicating rule system 'crs' is defined by the parallel composition of a set of rule systems

$$RS(crs) := \{rs_1, \ldots, rs_k\}.$$

The set of possible states Q_{crs} of a communicating rule system is given by the cartesian product of the states of its components as

$$Q_{crs} := Q_{rs_1} \times \ldots \times Q_{rs_k}.$$

A typical state is $q_{crs} = (q_{rs_1}, \ldots, q_{rs_k})$.

Example:
The set of rule systems composed in parallel in the rule system Network_service is:

```
RS(Network_service)   = {Network_user1, Network_user2,
                         Network_entity1, Network_entity2,
                         Network}
```

The following examples illustrate how states from $Q_{Network\_entity1}$ are formed. However, only some excerpts of the whole cartesian product can be shown:

```
qNetwork_entity1^0 = ((r1, p1, d1, 0, 0, 0, 0, ...), ({}))
qNetwork_entity1^1 =
   ((r1, p1, d1, 0, 0, 0, 0, ...),
    ({(n_con_req,
      ({"called", "calling", FALSE, TRUE, {...}, "hello"},
       {{FALSE, FALSE, modulo_8, 0, 1}, '00001011'B, 5, 5, "12345", "67890", 0, "hello"},
       ...
      ) ) } ) )
qNetwork_entity1^2 = ((r1, p2, d1, 0, 0, 0, 0, ...), ({}))
```

The three examples above refer to states of the rule system Network_entity1. The rule n_con_req can be started in the state $q_{Network\_entity1}{}^0$, n_con_req has been started and is running in the state $q_{Network\_entity1}{}^1$, and it has been ended and cannot be started in the state $q_{Network\_entity1}{}^2$. These examples are used later in this section, when the requirements for starting and ending rules are formalized.

The following examples illustrate how states from $Q_{Network\_user1}$ are formed. Since the specification of the rule system Network_user1 was not given in section 3.2, the cartesian product is not detailed. However, it is assumed that the rule establish_connection is running in the state $q_{Network\_user1}{}^1$ and that it is no longer running in the state $q_{Network\_user1}{}^2$.

```
qNetwork_user1^1   = ((...), ({(establish_connection, (...))}))
qNetwork_user1^2   = ((...), ({}))
```

The following examples illustrate how states from $Q_{Network\_service}$ are formed:

```
qNetwork_service^1 = (qNetwork_user1^1, ..., qNetwork_entity1^0, ..., ...)
qNetwork_service^2 = (qNetwork_user1^2, ..., qNetwork_entity1^1, ..., ...)
```

Actions

According to the classification, a basic rule system cannot use gate instances as state objects, but only request or offer events at gates given to the basic rule system as actual parameters in the instantiation. The gates given as parameters to a rule system instance form the interface of this rule system to the environment for both basic and communicating rule systems. The events occurring at those gates are **external events**, perceptible by an external observer. In communicating rule systems, gate instances may be used as state objects to be shared by rule system instances. The events occurring at those gates are **internal events**, not visible for an external observer. Thus, two sets of gates are derived from the STATE-part of the specification of a rule system 'rs':

- Ext_Gates(rs), containing all gates declared as formal parameters (external gates);
- Int_Gates(rs), containing all gates declared as state objects (internal gates).

The set of gates of a rule system 'rs' and the overall set of gates in a CRS specification is defined as

$$\text{GATES(rs)} := \text{Ext\_Gates(rs)} \cup \text{Int\_Gates(rs)}$$

$$\text{GATES} := \bigcup_{i\in\{1,\ldots,k\}} \text{GATES}(rs_i).$$

With each g∈GATES, a finite set ETYP(g) of valid event types is associated having the admissible parameter domains EPAR(g,etyp).

```
Example: (the gate specifications are given in Figure 54)
Ext_gates(Network_entity1) = {NSAP1, Packets1}
Int_gates(Network_entity1) = {}
Ext_gates(Network_service) = {}
Int_gates(Network_service) = {NSAP1, NSAP2, Packets1, Packets2}
GATES(Network_entity1)     = {NSAP1, Packets1}
GATES                      = {NSAP1, NSAP2, Packets1, Packets2}
ETYP(NSAP1)                = {N_CONNECT_req, N_CONNECT_ind, ...}
EPAR(NSAP1, N_CONNECT_req) = N_CONNECT_request (see Figure 38)
```

From the gate specifications the set of actions is derived as:

Definition 3.3.2.a

EVENTS := { (g, etyp, epar) | g∈GATES ∪ {I},
etyp∈ETYP(g) ∪ {*internal*},
epar∈EPAR(g, etyp)}

EVENT_CLASSES := {*request, offer, event*}

ACTIONS := { (e, eclass) | e∈EVENTS, eclass∈EVENT_CLASSES}

The events occurring at a gate 'g' have to be of an event type 'etyp' that has been defined for 'g'; events may convey parameters 'epar' from the parameter domain determined by 'etyp'. An event without parameters is treated as a special case with an empty parameter domain. Each event either represents an *offer*, *request*, or *event*. The set of events includes the **internal event**, not visible to the external environment. The symbol **I** is used to denote an anonymous gate for the *internal* event type with no parameters.

An element $a \in$ ACTIONS is either of the form

$$a = (a_R, \textit{request}), \quad \text{or} \quad a = (a_O, \textit{offer}), \quad \text{or} \quad a = (a_E, \textit{event}) .$$

```
Example: (see Figure 37 and Figure 38 for the type definitions)

e¹  = (NSAP1, N_CONNECT_req, ({"called", "calling", FALSE, TRUE, {...}, "hello"}))
e²  = (Packets1, Call_request, ({{FALSE, FALSE, modulo_8, 0, 1},
                                  '00001011'B, 5, 5, "12345", "67890", 0, "hello"}))

a_R¹ = e¹
a_O¹ = e²
a_E¹ = e¹

a⁰  = (a_E¹, event)
a¹  = (a_R¹, request)
a²  = (a_O¹, offer)
```

Transitions (startable and endable rules)

In CRS, rule systems communicate via events at gates. This form of communication is controlled by the rules in a rule system. Rules may request events in their start condition and offer events in the effects formula.

Let

requests(rs, r)

be the set of event requests (possibly empty) belonging to the $F_{cond\,(rs,\,r)}$ formula of a rule 'r' in a rule system 'rs'. Each event request is of the form req = (g, etyp, $\underline{lv}$(req))

where

g ∈ GATES(rs),
etyp ∈ ETYP(g),
$\underline{lv}$(req) is a tuple of local variables in LV(rs) denoting the requested event parameters.

Similarly, let

offers(rs, r)

be the set of event offers (possibly empty) in the $F_{effects\ (rs,\ r)}$ formula of a rule 'r' in a rule system 'rs', with elements

off = (g, etyp, $\underline{lv}$(off))

where

g ∈ GATES(rs),
etyp ∈ ETYP(g),
$\underline{lv}$(off) is a tuple of local variables in LV (rs) denoting the offered event parameters.

For a logical evaluation, the requests and offers in rule start condition and effects formulas have to be regarded as propositional variables. To determine the truth value of the whole formula, an assignment function

$$S: SV(rs) \times LV(rs) \times requests(rs, r) \times offers(rs, r) \rightarrow$$
$$StatVal(rs) \times LocVal(rs) \times \{TRUE, FALSE\} \times \{TRUE, FALSE\}$$

is required. This assignment reflects the state of the rule system (state variables), the execution state (local variables) of the rule, and information about occurring events.

```
Example:
requests(Network_entity1, n_con_req) = {(NSAP1, N_CONNECT_req, (N_CR_SP))}

offers  (Network_entity1, n_con_req) = {(Packets1, Call_request, (CR_PDU))}
```

The SYN-part of a rule system specification delivers two irreflexive relations, called *compatibility* and *priority*:

$$\textbf{comp}_{rs} \subseteq \text{RULES(rs)} \times \text{RULES(rs)}$$

$$\textbf{prior}_{rs} \subseteq \text{RULES(rs)} \times \text{RULES(rs)}\ .$$

Compatibility of two rules is expressed by using the parallel (||) and semi-parallel (|-) operators. Priority of a rule is defined by using the prior ($\prec$) operator. If the SYN-part is empty, so are the relations **comp** and **prior**. Thus, all rules are mutually exclusive and the choice which rule to start among those that are startable with respect to state and communication requirements is nondeterministic.

```
Example: (see the specification example given in Figure 51)

comp_Network_entity1  = {(fragment_data, send_data),
                         (send_data, fragment_data),
                         (send_data, receive_data),
                         (receive_data, send_data),
                         (receive_data, reassemble_data),
                         (reassemble_data, receive_data)}

prior_Network_entity1 = {(send_expedited_data, send_data),
                         (receive_expedited_data, receive_data)}
```

For starting a rule 'r' in a basic rule system 'rs', the start condition of 'r' has to be satisfied. Let 'actoffs' $\subseteq$ EVENTS be a set of actual event offers which are mapped by an injective function 'φ' to some or all of the event requests of the rule 'r' (those which evaluate to 'true' in $F_{cond\,(rs,\,r)}$). The requirements on the given state of 'rs' and given event requests are summarized in the following predicate:

Definition 3.3.2.b

startable(r, actoffs, q, $\underline{lv}$) :⇔

$F_{cond(rs, r)}(q) \wedge$

$(\exists \varphi: \varphi: actoffs \rightarrow requests(rs, r)$

$\wedge\ \varphi$ injective

$\wedge\ \forall off=(g_1, etyp_1, epar) \in actoffs$

$\forall \varphi(off)=(g_2, etyp_2, \underline{lv}(\varphi(off))):$

$(g_2=g_1 \wedge etyp_2=etyp_1 \wedge \underline{lv}(\varphi(off)) = epar))$

where rs is a basic rule system, r∈RULES(rs), q∈Q_{rs}, actoffs ⊆ EVENTS.

The predicate ***startable*** expresses that in order to start a rule 'r' in a rule system 'rs' in a given state 'q' with actual event offers 'actoffs', the start condition must be satisfied and the actual event offers have to match the event requests of the rule 'r'. The values of the local variables of the rule 'r' must be equal to the values of the offered event parameters.

In a similar way, the following predicate summarizes the requirements for ending a rule:

Definition 3.3.2.c

endable(r, actreqs, 'q, q, '$\underline{lv}$, $\underline{lv}$) :⇔

$F_{effects(rs, r)}('q, q) \wedge$

$(\exists \varphi: \varphi: actreqs \rightarrow offers(rs, r)$

$\wedge\ \varphi$ injective

$\wedge\ \forall req=(g_1, etyp_1, epar) \in actreqs$

$\forall \varphi(req)=(g_2, etyp_2, \underline{lv}(\varphi(req))):$

$(g_2=g_1 \wedge etyp_2=etyp_1 \wedge \underline{lv}(\varphi(req)) = epar))$

$\wedge$ (r, '$\underline{lv}$)∈'rrlv

where rs is a basic rule system, r∈RULES(rs), 'q=('$\underline{sv}$, 'rrlv), q=($\underline{sv}$, rrlv)∈Q_{rs}, actreqs ⊆ EVENTS.

The predicate ***endable*** expresses that in order to end a rule 'r' in a rule system 'rs' in a given state ''q' with actual event requests 'actreqs', resulting in the new state 'q', the effects formula must be satisfied and the actual event requests have to match the event offers of the rule 'r'. The values of the local variables of the rule 'r' must be equal to the values of the requested event parameters. In addition, the rule 'r' must be running in the state ''q'.

Example:
The predicates

<u>startable</u>(n_con_req, $\{a_R{}^1\}$, $q_{Network\_entity1}{}^0$, (N_CR_SP))

and

<u>endable</u> (n_con_req, $\{a_O{}^1\}$, $q_{Network\_entity1}{}^1$, $q_{Network\_entity1}{}^2$, (N_CR_SP), (CR_PDU))

are satisfied.

3.3.3 Interleaving Semantics

In this section, the definitions of the LTSs for basic and communicating rule systems are given based on the notions defined in the previous section. First, a basic rule system is considered in isolation and its behavior is modelled by a so-called "local" LTS. Then, the "local" LTSs are used to compose the "distributed" LTS of a communicating rule system.

Recall from Definition 3.3.1.a that an LTS is defined by a set of states, a set of actions, and a transition relation between states labelled with actions. Since rule systems are based on the EFSM model, the state space of the LTS is defined by the state objects. The transition relation is determined by the rules controlling internal and external actions of rule systems.

As far as the action set of the LTSs is concerned, the definition is not so straightforward. Generally speaking, the elements of the action set are the *atomic observations* that can be seen when observing the system and the transition relation determines the possible temporal orderings of these observations. Choosing different forms of atomic observations has a large impact on the semantics of concurrency [Acet87].

In CCS [Miln80] and CSP [Hoar85], single events are regarded as the atomic observations and the resulting semantical model is called **interleaving semantics**. Concurrency between systems performing a set of actions is expressed by allowing actions to occur in any total order. This view suffices to model the behavior of most concurrent systems.

However, some authors emphasize that parallelism is more than arbitrary interleaving and consider sets of actions the atomic observations. Accordingly, the semantical model is called **multiset semantics**.

In [Acet87], even multiset semantics is criticized and an enhanced model is proposed by defining a partial order on the elements of a set of actions in order to express causal independency (**partial ordering semantics**).

The semantics for CRS defined in this section is based on the interleaving model. This is because interleaving semantics is considered to be sufficient for protocol engineering tasks to be supported in an integrated tools environment as presented in chapter 5 of this thesis.

The tool support is based on operational formal descriptions that are executed by runtime mechanisms in the kernel of the environment. These runtime mechanisms are more easy to develop and they work more efficiently if interleaving semantics is assumed. A multiset semantics for CRS has been defined in [Schn90a].

Semantics of basic rule systems

Using the auxiliary definitions given in section 3.3.2, the behavior of a basic rule system 'rs' without rule priorities (WP) is described by the following LTS:

Definition 3.3.3.a

The labelled transition system LTS_{rs}^{WP} for a basic rule system 'rs' without priorities is defined as

$LTS_{rs}^{WP} = (Q_{rs}, ACT_{rs}, -\to_{rs}^{WP})$, where

Q_{rs} := StatVal(rs) × RunRulesLocVal(rs)

ACT_{rs} := ACTIONS

$-\to_{rs}^{WP} \subseteq Q_{rs} \times ACT_{rs} \times Q_{rs}$ is given by

$'q_{rs} \; -a\to_{rs}^{WP} \; q_{rs}$:⇔
(∃ a_R, actoff:
(a = (a_R, *request*) ∧ actoff = {a_R} ∧
∃ startrule∈RULES(rs)\RunRules(rs): ∃ $\underline{lv}_{startrule}$:
((***startable***(startrule, actoff, $'q_{rs}$, $\underline{lv}_{startrule}$)
∧ ∀ r'∈RunRules(rs): ***comp***$_{rs}$(startrule, r'))
∧ rrlv = 'rrlv ⋃ {(startrule, $\underline{lv}_{startrule}$)})))
∨
(∃ a_O, actreq:
(a = (a_O, *offer*) ∧ actreq = {a_O} ∧
∃ endrule∈RunRules(rs): ∃ $\underline{lv}_{endrule}$: ∃ '$\underline{lv}_{endrule}$:
(***endable***(endrule, actreq, $'q_{rs}$, q_{rs}, '$\underline{lv}_{endrule}$, $\underline{lv}_{endrule}$)
∧ rrlv = 'rrlv \ {(endrule, '$\underline{lv}_{endrule}$)})))

where q_{rs} = ($\underline{sv}$, rrlv), $'q_{rs}$ = ('$\underline{sv}$, 'rrlv).

A step in LTS_{rs}^{WP} is either a rule start or a rule end. An action of LTS_{rs}^{WP} is either an event request (a_R) of a rule that is started ('startrule'), or an event offer (a_O) of a rule that is ended ('endrule'). In either case, there is only a single event requested or offered, respectively. In case of a rule start without an event request or a rule end without an event offer, the action is the internal action **i** with event type *internal* and no parameters. Rule starts and ends are totally ordered and cannot occur simultaneously. If several rules satisfy the requirements for a rule start or end, one rule is selected in a nondeterministic fashion.

```
Example:
The transitions
```

$$q_{Network\_entity1}{}^{0} -(a^1)\rightarrow_{Network\_entity1}{}^{WP}\ q_{Network\_entity1}{}^{1}$$

and

$$q_{Network\_entity1}{}^{1} -(a^2)\rightarrow_{Network\_entity1}{}^{WP}\ q_{Network\_entity1}{}^{2}$$

are possible steps of $LTS_{Network\_entity1}{}^{WP}$.

The degree of nondeterminism for starting rules is restricted if priorities between rules are included. This is done by the following definition:

Definition 3.3.3.b

The labelled transition system LTS_{rs} for a basic rule system 'rs' is defined as

$$LTS_{rs} = (\ Q_{rs},\ ACT_{rs},\ -\rightarrow_{rs}\ ),\ \text{where}$$

Q_{rs} := StatVal(rs) × RunRulesLocVal(rs)

ACT_{rs} := ACTIONS

$-\rightarrow_{rs}\ \subseteq Q_{rs} \times ACT_{rs} \times Q_{rs}$ is given by

$$'q_{rs} \ -a\rightarrow_{rs}\ q_{rs} \quad :\Leftrightarrow$$
$$(\ 'q_{rs}\ -a\rightarrow_{rs}{}^{WP}\ q_{rs}\ \wedge$$
$$(\ a=(a_R,\ \mathit{request}) \Rightarrow$$
$$(\forall\ a_R':\ ('q_{rs}\ -(a_R',\ \mathit{request})\rightarrow_{rs}{}^{WP}\ q_{rs} \Rightarrow$$
$$\neg\ \mathbf{\mathit{prior}}_{rs}(startrule_{a_R'},\ startrule_{a_R}))))$$

This definition makes use of the previous one by stating that a transition of LTS_{rs} must be a possible transition of $LTS_{rs}{}^{WP}$ as well. However, the following restriction is defined: if the transition action 'a' represents an event request 'a_R' of a rule that is started ('$startrule_{a_R}$'), then all other rules with the same or another event request 'a_R'' that could have been started as well ('startrule$_{a_R'}$') must have the same or lower, but no higher priority.

Example:
With the specification example of Figure 51, a transition of $LTS_{Network\_entity1}$ with rule send_data as the 'startrule' is not possible if a transition with rule send_expedited_data as the 'startrule' is possible as well.

Semantics of Communicating Rule Systems

The LTSs for basic rule systems (LTS_{rs}) are used to compose the LTS of a communicating rule system (LTS_{crs}) as in the following definition:

Definition 3.3.3.c

The labelled transition system LTS_{crs} for a communicating rule system 'crs' is defined as

$$LTS_{crs} = (Q_{crs}, ACT_{crs}, -\to_{crs}), \text{ where}$$

$$Q_{crs} := Q_{rs_1} \times \ldots \times Q_{rs_k}$$

$$ACT_{crs} := ACTIONS$$

$$-\to_{crs} \subseteq Q_{crs} \times ACT_{crs} \times Q_{crs} \text{ is given by}$$

$$'q_{crs} -a\to_{crs} q_{crs} \quad :\Leftrightarrow$$
$$(\exists a_R=(g_R, etyp_R, epar_R): a = (a_R, \mathit{request}) \wedge g_R \in Ext\_gates(crs)$$
$$\wedge \exists rs \in RS(crs): 'q_{rs} -(a_R, \mathit{request})\to_{rs} q_{rs})$$
$$\vee$$
$$(\exists a_O=(g_O, etyp_O, epar_O): a = (a_O, \mathit{offer}) \wedge g_O \in Ext\_gates(crs)$$
$$\wedge \exists rs \in RS(crs): 'q_{rs} -(a_O, \mathit{offer})\to_{rs} q_{rs})$$
$$\vee$$
$$(\exists a_E=(g_E, etyp_E, epar_E): a = (a_E, \mathit{event}) \wedge g_E \in Int\_gates(crs)$$
$$\wedge \exists rs_{req} \in RS(crs): 'q_{rs_{req}} -(a_E, \mathit{request})\to_{rs_{req}} q_{rs_{req}}$$
$$\wedge \exists rs_{off} \in RS(crs): 'q_{rs_{off}} -(a_E, \mathit{offer})\to_{rs_{off}} q_{rs_{off}}$$
$$\wedge rs_{req} \neq rs_{off})$$

where $q_{crs} = (q_{rs_1}, \ldots, q_{rs_k})$.

The "distributed" LTS_{crs} of the communicating rule system 'crs' may perform a transition with an event request (a_R) or offer (a_O) at one of the external gates (event classes *request* and *offer*) as the transition action if there is a component rule system 'rs' making the request or offer during a transition of its "local" LTS (first two cases in the definition). If the transition action is of the class *event*, there have to be two distinct component rule systems 'rs_{req}' and 'rs_{off}' performing transitions with the same event (a_E) tagged as *request* for 'rs_{req}' and as *offer* for 'rs_{off}'. These rule systems participate in a two-way rendezvous. The behavior of the communicating rule system 'crs' is described by the possible transition sequences of LTS_{crs}.

Example:
With rs_{req} = Network_entity1, rs_{off} = Network_user1, the transition

$$q_{Network\_service}{}^{1} \ -(a^{0})\rightarrow_{Network\_service} \ q_{Network\_service}{}^{2}$$

is a possible step of $LTS_{Network\_service}$.

Using CRS semantics

The benefits of CRS semantics being defined in terms of LTSs can be summarized as follows:

1. The definitions provide an unambiguous reference for the meaning of rule system, rule, and gate specifications. The possible behavior of rule systems and the communication at gates is precisely defined.

2. LTSs are widely accepted as a basic model of concurrent systems. The notions of states and actions can be used to verify desired properties of systems being described. Systems can be compared with respect to their external observability to see whether equivalence relations or preorder relations hold [DeNi87]. Equivalences and preorders are the mathematical means to describe the relation between a specification and an implementation in a formal way [Brin90].

3. The possible set of states and actions and the definition of the transition relation of an LTS provides complete information for building specification interpreters and runtime environments for the simulated execution of formal descriptions. The executability of the specification can be exploited in various ways by tools supporting the system development life-cycle.

The focus of this thesis is on the last item of this list. A runtime environment for the CRS language, a way of achieving operational specifications, and an integrated tools environment for developing communication systems is described in chapter 5.

4 Protocol Engineering with Formal Description Techniques

The development of communication systems is a complex task requiring considerable effort in each step from design to implementation and testing. An outline of a development process based on a formal methodology was given in chapter 1. During development, the initial requirements, concrete design, system refinements, implementation, and tests of communication services and protocols have to be described and documented. In case of communication in open systems, the requirements are defined by international standards. The benefit of using FDTs in order to achieve precise and unambiguous specifications has been recognized by the standard committees. The potentials of FDTs for automating development steps by employing tools to process the specifications is interesting from an implementor's point of view.

In particular, executable specifications can be employed at several stages in the development process and specification interpreters often form the heart of protocol engineering environments. How executability of specifications is achieved for some existing techniques is outlined in this chapter and some existing software environments supporting protocol engineering tasks based on these techniques are described.

4.1 Executable FDTs

During the development of a protocol, executable specifications can be employed for various purposes ranging from design validation and checking of refinements to the generation of implementation code frames and test cases. Executable specifications still contain all implementation options and may exhibit nondeterministic behavior, whereas in an implementation, the options have been resolved in a deterministic fashion. Depending on the FDT, complex runtime support has to be provided for executing specifications. This is investigated for some existing techniques in the following.

The execution of FDTs is based on their individual semantic models. The single steps and possible execution sequences of specified systems are given in terms of the primitives of these semantic models (state transitions and reachability of states, transition occurrences and reachability of markings, event occurrences and observable behavior, evaluation and truth values of formulas etc.). Depending on the semantic model used, executable specifications always run in specialized environments that offer idealized conditions (synchronous events, no loss of messages, unlimited buffer space, view on global states, discrete timing etc.). Depending on the degree of abstraction provided by a particular technique, runtime support has to be realized in order to provide these idealized conditions. Besides affecting the complexity of this runtime support, the semantic models of FDTs sometimes have a dedicated paradigm, or style of programming, that is favorable for achieving executable specifications.

4.1.1 Estelle, SDL, and PASS

Due to the underlying EFSM model and the use of programming language-like constructs, the compilation of formal descriptions written in Estelle or SDL into procedural code is feasible. In general, Estelle and SDL specifications are close to implementations. The Estelle compilers described in [Boch87b] and [Vuon88] generate "PASCAL" and "C" code, respectively, employing a similar strategy. This strategy is suitable for techniques based on the EFSM model in general and can be summarized as follows:

- For each module specification, a procedure in the target programming language is generated that can be compiled separately (as an independent implementation module if the target language supports modularization). The procedure comprises code for checking transition conditions and for executing transition actions.

- The execution state of a module instance is represented by a record data structure, called a module *control block* (or module *descriptor*), storing the values of all state variables of the module instance. This control block is passed together with an input interaction and interaction parameters to the module procedure checking for enabled transitions (dummy interactions are used to trigger spontaneous transitions). The type definitions for module state variables, channels, interactions, and interaction parameters are generated by the compiler.

- The selection among module instances with enabled transitions is done by an overall transition scheduler according to some predefined scheduling strategy. For example, a round-robin strategy is employed by the compiler described in [Boch87b], but other strategies are possible as well. The tool described in [Vuon88] generates a tree data structure reflecting the module hierarchy as the central data structure used by the transition scheduler in order to preserve the semantics of the different forms of parallelism in Estelle.

- The selection among enabled transitions of a module instance is done by the module procedure being coded using embedded "case"- or "if"-statements. It is important that the compilers resolve the nondeterminism in the specification by taking some fixed strategy (for example, the textual order of transitions as defined in the module specification [Vuon88]).

- As part of the module control block, a queue of input interactions is maintained in FIFO order for each interaction point of the module (or one common queue for all interaction points). The communication structure is represented by using pointers to the control blocks of other module instances.

- The module control blocks are created at runtime by corresponding routines and are updated after each transition. If a module instance is released its control block is deleted. Initialization routines establishing the static specification structure are generated from the initial transition specifications and are called prior to the dynamic execution of the specified system.

Other approaches to the execution of Estelle specifications (not using compilation) exist as well. An Estelle interpreter called ESTIM is described in [Saqu89]. The global execution model employed is fully consistent with the Estelle semantics defined in [ISO9074], but additionally a rendezvous mechanism is provided and parallelism is restricted. The "systemprocess" and "process" attributes of modules are not supported, only "systemactivity" or "activity" may be used. The specification is represented as an attributed syntax tree. During the initialization phase, the tree is traversed depth-first and all module instances are created. The global state obtained is represented as an 8-tuple of semantic functions that map module references to module headers and bodies, interaction points to connected or attached counterparts, variable references to memory locations etc. These functions are realized according to denotational semantics and are implemented using the functional programming language "ML".

An object-oriented model for Estelle is defined in [Sije89] and a corresponding simulator called WISE has been realized in "SMALLTALK". A hierarchy of classes and corresponding class behavior is defined for all possible Estelle objects. *Definition* classes represent type, procedure, and transition specifications, *declaration* classes are used to create variable and interaction instances, and active module instances and interaction points are created from *activation* classes. A particular system consists of a collection of instances of these classes. During dynamic execution of a specification, messages are exchanged between object instances to trigger transitions, update variables, add or remove interactions from queues etc.

Due to the implementation-orientation of Estelle and SDL specifications are interpreted at a low level of abstraction. The compilers for these techniques generate deterministic code, thereby loosing important information given in the specification. The Estelle interpreters are based on "meta-models" of Estelle semantics defined in the denotational and object-oriented style, respectively. This additional modelling allows the execution of Estelle specifications by functional and object-oriented interpreters. However, the relation to the transition system semantics of Estelle is not obvious [Sije89].

Various experiments on how to achieve executable specifications based on the EFSM model have also been conducted for the PASS (*Parallel Activities Specification Scheme*) technique [Flei87b] [Flei88] [Flei89]. In PASS, a graphical syntax similar to SDL is employed for specifying the state machine (PASS graph) of a protocol. A number of tools have been implemented that generate different forms of executable code from the PASS graph representation. In the "code-driven" form, the PASS graph is translated into procedural code with entries for each transition. In the "data-driven" form, the PASS graph is mapped onto a data structure interpreted at runtime. Surprisingly, no efficiency is lost in the "data-driven" form, but the code size is reduced considerably. Alternatively, a set of "PROLOG" clauses can be generated from the PASS graph as well. The resulting "PROLOG" program represents an executable version of a PASS specification. Unfortunately, PASS specifications only capture the state machine aspects (no data aspects) and only allow deterministic behavior.

4.1.2 Petri Nets

A large number of tools for the simulation and animation of Petri nets exist today based on different classes and variants of nets. An overview is given in [Feld87]. The simulators developed for the PROTEAN system (see [Bill88]) and the one used in the PROSIT project (see [Paul85]) are described briefly, since they are applied to protocols.

PROTEAN is based on *Numerical Petri Nets* (see section 2.2.2), whereas *Product Nets* have been used in PROSIT. Both net classes are variants of high-level Petri nets as defined in chapter 2. The net simulators take a net specification in a textual form as input and, after static analysis, generate "PASCAL" code that mainly comprises the following parts:

- data structures representing the static net structure (places and transitions);
- procedures checking transition conditions and executing transition actions;
- type definitions for token colors and data variables;
- initialization code for establishing the initial marking.

In the PROSIT simulator, places are identified with integer numbers and a pointer array is used for referencing token objects present at places during runtime. All token objects that are created and deleted during runtime are kept in a global object pool. Transitions are represented by the numbers of connected places, pointers to token objects that are moved when the transition occurs, and information about the type and number of tokens. The net markings are stored in a three-dimensional table holding the number of the marking, the actual object pool, and object pointers for each place.

After the net specification has been compiled, the net components can be displayed and edited using a graphical editor in the PROTEAN system. The specification objects appear one after the other and can be adjusted on the screen by the user of the tool. When starting a simulation, the initialization code is executed and the initial marking of the net is displayed. Zooming functions are available to focus on specific parts of the net. All transition conditions are evaluated and the set of enabled transitions is displayed (either graphically by highlighting or using color, or textually in form of a list). The user of the tool selects an enabled transition and the transition actions are executed, resulting in a new marking. The net markings reached during net simulation are saved for backtracking purposes. The whole, or selected parts of the reachability graph of a net can be generated this way.

For the interactive use of Petri net simulators sophisticated graphical user interfaces are required. In an automatic mode, nets can be thoroughly analyzed with respect to various properties (reachability, invariant conditions etc.). The structure of data objects in a net specification is, however, rather simple in most tools. As a consequence, the net simulation is done at a low level of granularity. Generating the reachability graph for protocols of realistic size causes both space and efficiency problems.

4.1.3 LOTOS

Interpreters for LOTOS are described in [Logr88] (ISLA) and in [Tret89] [Eijk89b] (HIPPO). The interpretation is based on term rewriting techniques for both the behavior and the data part of LOTOS. The semantics of LOTOS has been defined using *labelled transition systems*, where the transitions describe the transformation of behavior expressions after performing interactions. Abstract data types are defined in LOTOS using the algebraic specification language ACT ONE. Consequently, the semantics of both parts of the LOTOS language can be directly implemented as rewrite rules.

In the ISLA simulator, the rewrite rules for behavior expressions are implemented as "PROLOG" rules (see [Logr88]). As an example, the semantics of the *choice*-operator in LOTOS is expressed as a "PROLOG" rule in Figure 63. The interpreter runs on the internal "PROLOG" representation implementing the dynamic semantics of LOTOS as defined in [ISO8807]. Whenever a value expression has to be evaluated, an ADT interpreter is called as a subordinate routine, using a "leftmost-innermost" evaluation strategy. The rewrite

```
Let 'B', 'B1', and 'B2' be behavior expressions, 'a' an interaction.
The semantics of the choice-operator is given as follows:

   if B1 -a→ B2 then B1 [] B -a→ B2 .

This is implemented in "PROLOG" as:

   infer( choice( B1, B ), a, B2 ) :- isaction( a ), infer( B1, a, B2 ) .
```

Figure 63. Example of LOTOS semantics expressed as "PROLOG" rules

rules for the ADTs in a specification are generated by a front-end tool after parsing the specification and checking the static semantics.

The HIPPO simulator (see [Eijk89b] [Tret89]) also uses term rewriting to compute possible continuation alternatives from a given behavior expression as the result of a 'menu' function. The user of the tool has to make a selection and has to supply interaction parameters before invoking the 'next' function. LOTOS semantics in the form of *communication trees* are used as the underlying formalism (see [Eijk88]). The tool has been implemented in "C".

The interpretation of LOTOS specifications based on term rewriting works well provided that the derived rewrite systems are *confluent* and *terminating.* Confluent means that the choice among applicable rewrite rules and the order in which rewrite rules are applied do not affect the resulting behavior. Terminating means that all computations have to be finite. Non-confluent rewrite systems can be transformed into confluent ones by applying the "Knuth-Bendix Algorithm", which is offered as an option in the ISLA simulator. The termination property of a given rewrite system is not decidable in general [Joua87].

An alternative approach to the execution of LOTOS specifications is described in [Gilb89] [Gilb87]. A "source-to-source translation" from LOTOS to "PARLOG" (a parallel logic programming language) is proposed. However, only the behavior part of LOTOS is translated and "PARLOG" typing mechanisms are used for the data part. Even for the behavior part, a direct translation could not be achieved due to the different semantics of the choice and parallel operators of LOTOS and "PARLOG". Instead, parameterized

"PARLOG" programs have been designed to behave like LOTOS processes and gates.

Some work has been started recently on transforming LOTOS processes into state machines [Eijk90b] [Mana89]. The approach has been theoretically treated in [Karj88]. A prerequisite to accomplishing the transformation is that the input specification is close enough to an implementation. Ideally, the specification has been written in a *state-oriented specification style* [Eijk90b]. The behavior part and the data part are kept separate. The tool described in [Mana89] generates a finite automaton for each behavioral unit in the specification that synchronizes with others. Each LOTOS process is compiled into a reentrant "C" function that is executed as a separate operating system process. A complex runtime kernel using system calls to determine possible multi-way synchronizations is required. In the approach described in [Eijk90b] this was avoided for efficiency reasons. Instead, a given LOTOS specification is transformed by a tool implementing the LOTOS *expansion theorem*. By applying this theorem, parallelism (and thus the need for communication) is eliminated by introducing nondeterminism. For the data part, the tool described in [Mana89] derives a rewrite system. In the implementation experiment reported in [Eijk90b], ADTs have been translated manually.

Summarizing, the complete translation of LOTOS into an executable language is difficult to achieve. The LOTOS transition system semantics can be implemented directly as rewrite systems that are executed by interpreters. The execution of specifications is performed on an abstract level by manipulating symbols. However, various implementation problems have to be solved: termination of ADT rewrite systems, determination of multi-way synchronization, value generation in interactions, and process disruption. The interpretation mechanisms have to cope with efficiency problems for specifications of realistic sizes.

4.1.4 Logic interpreters

Logic interpreters execute specifications given in the form of logic programs. Several experiments have been reported in the literature using "PROLOG" as a tool for executing logic specifications of communication services and protocols (see for example [Cleg89] [Boch86] [Sidh83]). In

"PROLOG", logic programs are given as a set of "Horn clauses" (see [Cloc81]) that may have been obtained from a more general logic specification by translation. For example, in the specification environment for the Z notation described in [Mast89] specifications are translated into a set of "PROLOG" clauses to form the "domain-specific knowledge base" of the system.

Confronted with a "goal clause" of the form

$\leftarrow C$

the "PROLOG" interpreter tries to reason backwards from the goal to possible solutions. It searches a clause for which the head of the clause matches the goal C, normally consisting of a predicate symbol and arguments in form of constants, variables, or terms (for example, $C = P(t_1, \ldots, t_m)$). A more concrete example is given later. A matching clause

$B \leftarrow A_1, \ldots, A_n$

is found if either B is equal to C, or if B and C can be *unified* by applying a variable *substitution* to both. In such a substitution, variables are bound to terms that may be composed of constants, function symbols, and variables. All variable occurrences in B and C are substituted by the corresponding terms. If θ denotes a substitution (a set of variable replacement rules), and $B\theta$ its application to B, $C\theta$ its application to C, then

θ is called a *unifier* for B and C $\Leftrightarrow$ $B\theta = C\theta$.

If a matching clause can be found in the logic program, the *resolution* inference rule is applied and the goal C is reduced to the subgoals $(A_1, \ldots, A_n)\theta$. Strictly speaking, the problem of solving the goal C is reduced to the problem of solving B after applying θ to both. Then, the goal is to solve $B\theta$ by solving the subgoals $A_1\theta, \ldots, A_n\theta$. The computation stops, if matching clauses of the form

$B \leftarrow$

have been found for all subgoals. The "PROLOG" interpreter tries to solve subgoals from left to right (resolving "AND"-nondeterminism) and searches for matching clauses in the order given in the logic program (resolving "OR"-nondeterminism). Whereas the order of solving subgoals (*computa-*

tion rule) has no impact on the number and results of successful computations, the selection of matching clauses is important. If the computation does not succeed with a selected clause, backtracking is performed and the next matching clause is taken. This is done in a depth-first manner and may lead to infinite loops if recursion on the same clause occurs over and over again. Breadth-first search was not implemented because of enormous space requirements. The whole computation is performed on a symbolic level. Unification, the most time-consuming step, relies on pattern matching algorithms.

A procedural interpretation of "PROLOG" was given in [Kowa79] and may serve as a complementary explanation of the "PROLOG" interpretation mechanism. The clause

$B \leftarrow A_1, \ldots, A_n$

is interpreted like a procedure

```
procedure B
   begin
      call A1,
      call A2,
        ...
      call An
   end .
```

The goal clause

$\leftarrow C$

is interpreted like a procedure call and unification is used to pass parameters either by value (variables substituted by ground terms), or by reference (variables substituted by variables). This view emphasizes the fact that besides knowing whether computations can be terminated successfully, the programmer is particularly interested in variable bindings established during the execution of the logic program.

As an example, a simplified "PROLOG" clause for describing one transition of an X.25 DTE is given:

```
transition( p1, p2, N_CONNECT_req, Call_request) ←.
```

The state reached and the output made upon reception of an `N_CONNECT_req` event in the state `p1` can be obtained using the following goal clause:

```
← transition( p1, S, N_CONNECT_req, O)
```

where S and O are variables. The "PROLOG" system will find a successful computation, which states that there is a transition from the state p1 with the input N_CONNECT_req. Additionally, the system will find the answer substitution θ, where

$$\theta = \{ S \rightarrow p2 , O \rightarrow Call\_request\} .$$

Logic interpreters can be used for multiple purposes on the basis of a single logic program. This feature has been called **invertible execution** in [Boch86]. If the behavior of a protocol entity is specified as a logic program, different goal clauses can be used to get information about transitions. The focus can be on output and states, on states accepting a certain input, on subsequent states from a given initial state etc.

In order to use the knowledge about a protocol specified as a set of clauses, additional clauses are added to the logic program. According to [Kowa79], a "PROLOG" program can be separated into

- a **logic part** and
- a **control part**,

both given as Horn clauses. The logic part represents the knowledge about the problem domain, whereas the control part specifies how to use the knowledge for generating answer substitutions.

An example of a control clause is shown in Figure 64 ([Boch86]). The clause DoUntil can be used to compute possible sequences of state transitions from the Initial_state via a number of intermediate Next_states to the Final_state. The interpretation may be invoked with several goals: simu-

```
DoUntil( Initial_state, Final_state,
         [Input | Input_tail], [Output | Output_tail]) ←
   transition( Initial_state, Next_state, Input, Output),
   DoUntil( Next_state, Final_state, Input_tail, Output_tail) .

DoUntil( Initial_state, Final_state, [Input], [Output]) ←
   transition( Initial_state, Final_state, Input, Output) .
```

Figure 64. Example control clause in a logic program

lated execution (given an initial state and an input sequence), trace checking (given an actual state and an observed input/output sequence), or test sequence generation (input/output sequence from an initial state to a final state).

Logic interpreters are elegant tools for executing specifications with various intentions in mind. However, care must be taken in the control part of a logic program when providing clauses to achieve the intended goals because of their relationship to the clauses of the logic part representing the problem-specific knowledge. The efficiency of the interpretation depends on the efficiency of the pattern matching algorithms. Serious problems are caused by possible infinite computations. Several "non-logical" constructs have been invented in "PROLOG" ("cut" etc.) in order to avoid the exploration of infinite paths. The fact that problem-specific knowledge and control-oriented clauses are mixed in a logic program, and the inflexibility of the interpretation algorithm are considered drawbacks of "PROLOG" as a tool for executing specifications of communication services and protocols.

4.2 Protocol Engineering Environments

Software development environments combining protocol engineering tools are needed to apply a formal development methodology for communication services in practice. The development support that can be provided in such environments is outlined in this section and illustrated with some examples of existing tools.

4.2.1 Validation environments

The step from a specification of the informal requirements of a communication system to a formal description (***formalization***, see chapter 1) can be supported by textual or graphical editors. The formal specifications achieved this way can be validated with respect to various aspects. Validation methods have been classified as "simulative" and "exhaustive" in [Boch87a]. These methods can also be applied to validate refined specifications against their predecessors.

The main purpose of **simulation environments** for FDTs is to provide tools and runtime support for translating a given specification into an internal form and to facilitate the interactive validation of the specification by interpreting the internal representation. Tools like ESTIM [Saqu89], WISE [Sije89], VEDA [Jard88], PATE/PACO [Flei88], ISLA [Logr88], HIPPO [Tret89], PROTEAN [Bill88] are examples of simulation environments for FDTs.

For service specifications, the user of these tools checks whether the specified behavior reflects the informal requirements by watching interaction sequences and examining their accordance with local and global service constraints. Given a protocol specification, the simulated behavior of the protocol can be compared to the service specification. Particular transitions, states, encoding/decoding of PDUs and the mapping between PDU fields and SP parameters may be investigated. In an interactive simulation, a certain number of interesting situations can be explored to validate specific properties expected from the specification. In long lasting automatic simulations, the aim is at finding deadlocks, livelocks, or other kind of erroneous behavior. Such situations may be recognized during a simulation by ending up in some unexpected state, looping without progress, receiving inappropriate or corrupted messages etc. Although complete certainty about general properties of a specification cannot be reached, the confidence in the correctness of a design is increased.

The number of alternative paths and situations that may be explored in reasonable time during the simulation of a specification depends on the flexibility of the simulation tool. Facilities have to be provided for driving the simulation to areas of interest, for retrieving information in each state, and for entering data conveniently. In ESTIM and VEDA, a user interface in the form of a command interpreter is employed. It has been pointed out in [Jard88] that such an interface is not user-friendly enough. Only closed systems that do not interact with the environment (the user of the tool) can be simulated in VEDA. In PATE/PACO, an external procedure is called during execution of each input/output transition. For the purpose of validation this procedure receives input from and directs output to a human operator. In order to simulate systems sharing interaction points with the environment, so-called *user-driven transitions* can be selected in WISE. The window interface of the SMALLTALK system under which WISE was realized provides the appropriate means to display information about module instances that are distributed over a number of logical sites. Module instance *observers* are windows in which state variable values, enabled or delayed transitions, contents of interaction point queues, and other pieces of information about specification objects are displayed. The simulation may be driven by selecting transitions from menus or by invoking an automatic mode.

Interpreting LOTOS specifications with ISLA or HIPPO allows interactive checking and validation of possible interaction sequences. A simulation starts with an initial behavior expression which is expanded in subsequent steps. The set of interactions in which the simulated system is able to participate is displayed when calling the 'menu'-function. The user selects an alternative and invokes the 'next'-function. After the exploration of a certain execution path, the user may 'backtrack' to previous behavior expressions. However, the simulators may become difficult to use for large specifications, as reported in [Guil88]. Behavior expressions become lengthy (more than 2000 characters for a transport service specification) and data values represented as ACT ONE terms may comprise many symbols (about 400 characters for a transport SP). Besides being hard to read, it is difficult to enter such expressions interactively. However, users can build a library of shorthands for common values that are recognized by the system to simplify the interactions with the simulated system.

The PROTEAN tool allows stepwise simulation of net specifications using graphics. However, it may be difficult to get a view on the global state and keep track of all system-wide changes during transitions. Observing control and data flow is possible in the zoomed part of the net. The markings reached during net simulation can be resumed for backtracking.

Logic interpreters offer simulation capabilities at a symbolic level. Interpretation has to be invoked by goal clauses contained in the control part of the logic program. The flexibility in driving a protocol simulation depends on the design of the control part. The user interfaces provided by logic interpreters are mostly textual (without any graphic facilities) because there is no intuitive graphical representation for logic clauses.

In general, a substantial lack of support for handling data structures (PDUs, SPs) can be recognized in existing FDT simulators. An environment for designing and validating PDUs based on ASN.1 is described in [Cleg89], called ASNST. ASNST comprises several components (all implemented in "PROLOG") for defining ASN.1 types, constructing instances, checking instances against their type definition, and encoding/decoding instances. Using this tool, complete application layer PDUs can be defined and sample instances can be created and checked against the type definition. A 'sample editor' is provided which converts the internal tree representation into the external format and vice versa. ASNST was anticipated as a validation and testing tool to fill the existing gap in automated processing of data structures used in protocols and services. However, its use for implementation is restricted.

As opposed to simulation, **exhaustive validation environments** support protocol engineers in checking general properties of specifications. These properties include

- absence of deadlocks;
- termination or cyclic behavior;
- boundedness on resources;
- absence of livelocks.

A number of exhaustive validation environments have been built during recent years to process protocol specifications based on the FSM or EFSM model (special input formats are, however, often used instead of a particular FDT): PANDORA [Holz83], VALIRA [Vuon87], PAV [Sabn87], PATE/PACO [Flei88], PROSPEC [Chow88], SPANNER [Agga88], EDS [Chun88] [Blum83], PROTEAN [Bill88]. Several variants of *reachability analysis* are employed by the tools in these environments. Given an initial global state, all states that are reachable from this initial state are generated. During the analysis, states should be detected that cannot be reached or left anymore (*sources, sinks*), or states that are reached over and over again. These states may indicate deadlocks, livelocks, or other errors in the specification. The main problem with reachability analysis is that the number of states becomes large, or even infinite, if more than the basic FSM aspects of a protocol specification are considered (*state space explosion*). This problem severely restricts the usefulness of reachability analysis as a validation method. Several proposals have been made in the literature for modified reachability analysis methods (see for example [West87]) and have been implemented in some of the tool environments listed above. An overview is given in [Sari90].

The environment described in [Karj86] is based on CCS (for protocol behavior) and LISP (for protocol data structures) and offers the possibility of defining logic assertions (temporal logic) that are checked during validation runs of the tool. Other environments based on process algebras are described in [Papp87] and [Yemi83]. In [Papp87], a transformational approach has been taken to prove the *observation congruence* of service and protocol specifications by deriving a common normal form. The CUPID environment described in [Yemi83] supports "multiple concurrent protocol representations" and provides validation assistance with tools translating to different techniques and applying several different analysis methods.

For all these environments, successful applications to real world protocols have been reported. Besides many problems encountered in the use of these tools, many misinterpretations, errors, bottlenecks, and unexpected behavior of existing protocols have been detected by analyzing the protocol specifications.

4.2.2 Implementation environments

The compilers described in [Boch87b] and [Vuon88] (see section 4.1.1) are tools for generating implementation code from an Estelle specification. A comprehensive toolkit tailored towards the semi-automatic implementation of communication protocols is described in [Flei89] based on the PASS specification language. Complete implementations are, however, difficult to achieve. What can be derived from the specifications are the required state information, the conditions and actions of transitions, and the structure of interactions. The parts that have to be added to the generated code are the following (see also [Boch87b]):

- realization of interprocess communication;
- interfaces to the operating system;

- memory and buffer management;
- administration and scheduling of protocol layers.

These extensions have to be developed manually, but can be provided in runtime libraries that are used for several protocol implementations (see for example [Flei87a] [Ding90]). A typical example for a hand-coded part that is often done separately for each protocol are the encoding/decoding routines.

The main issues that have to be resolved during the compilation of protocol specifications are layer structuring (threads, processes, procedures etc.), communication between layers (interrupts, messages, procedure calls etc.) and nondeterminism (scheduling, transition ordering etc.). A comprehensive overview of implementation issues for OSI communication systems is provided in [Svob89]. As reported in [Vuon88], generated code contains fewer errors and allows cheaper and faster protocol implementation. The efficiency of manually-coded implementations, on the other hand, was not achieved.

In the EDS protocol engineering environment, implementation code can be generated after exhaustive validation of an Estelle specification [Sidh90b]. Most SDL based environments are mainly implementation environments (for example [Gras90] [Atle90] [Chen90] [Mizu88]). These environments provide graphical editors and implementation code generators for different programming languages ("C", "PASCAL", "CHILL", object-oriented "C"). The generated code can be compiled and executed by "simulators" provided in these environments. However, these "simulators" work on a rather low level of abstraction, similar to debuggers for programming languages. The approaches to implement protocols from LOTOS specifications are not so advanced yet. Some implementation experiments are reported in [Eijk90b], but compilers are still in an early prototype stage [Mana89].

Whereas for lower layer protocols in the OSI model the control aspects constitute the major part of an implementation, the situation changes in the upper layers. In the application layer, application entities exchange structured data objects. This data has to be encoded into a common transfer representation of the open system at the sending site and decoded at the receiving side. The automatic processing of data structure descriptions and the generation of encoding/decoding routines is desirable, but not sufficiently supported in the implementation environments mentioned above. The use of "PASCAL" data types is not appropriate for describing abstract data structures that have to be transferred over communication lines. Efficient implementation schemes for algebraic data type definitions are difficult to find. All the SDL-based tools mentioned above do not support automatic translation of the data part of SDL specifications.

Promising approaches to the automatic processing of data structure descriptions for communication systems are based on ASN.1 (excluding the macro facility). The ASN.1 compilers described in [Neuf90] [Beve89] [Gora88a] are capable of generating both implementation language data structures and routines converting from a local representation to transfer syntax and vice versa. For each ASN.1 type, a corresponding type definition in the target language is generated. However, the translation is by no means one-to-one due to those features of ASN.1 that cannot be found in common implementation languages (see chapter 3). In addition, the processing of the data according to the protocol (behavioral part) has to be implemented on the basis of the generated type definitions, which still demands considerable effort.

4.2.3 Testing environments

Since the final result of a protocol development is a running physical system, and not a symbolic system representation, a formal verification is not possible and tests have to be performed. Testing means applying a finite set of selected test cases derived from the specification to the implementation. These test cases may reveal errors, or increase confidence that the system behaves correctly if no errors are detected. Several kinds of testing efforts exist in the area of communication systems, including testing of functional correctness, interoperability, performance, robustness, etc.

In case of communication in open systems, protocol implementations have to be tested for accordance with protocol standards (**conformance testing**). Here, the goal is to reduce the probability that implementations are still unable to interwork, although they have been implemented according to the same standard. Since numerous implementations are developed by different providers, testing each implementation against all others is not feasible. Instead, each *implementation under test (IUT)* has to pass a number of test cases of a standardized *test suite*. The importance of international conformance testing efforts has been widely recognized as being crucial for achieving the objectives of OSI. The ISO and CCITT standards committees are currently jointly developing a common methodology and framework for conformance testing [ISO9646]. Surveys of this work can be found in [Rayn87] [Sari89] [Linn90].

In general, the different tasks that have to be addressed in protocol testing are related to

1. **test suite development**, concerning
   a. determination of test purposes,
   b. test derivation for each test purpose,
   c. test specification in some suitable notation;
2. and **test suite execution**, concerning
   a. test selection for some specific IUT,
   b. derivation of executable test cases,
   c. test execution and analysis.

For all these different tasks, tool support is desirable and is currently the subject of intensive research. A software environment for protocol testing would ideally combine tools supporting all the activities above.

Test suite development is still a manual activity requiring much effort. The development of standardized test suites for OSI protocols is currently conducted by a number of experts from different institutions. The development process is, however, quite slow and international acceptance is hard to achieve. Besides speeding up the development process, automated test suite development is desirable for at least two more reasons. Firstly, developing good test suites for communication protocols is a complex task due to the concurrent and distributed execution of a protocol by several entities and the various implementation options. Secondly, assuring correctness of test cases with respect to the informal protocol standard is difficult. By using formal descriptions, "quality" measures for test suites can be defined mathematically and can be related to the specifications. The test cases generated from formal descriptions according to formal methods are inherently correct with respect to the specification.

Generating *test sequences* from specifications based on an underlying FSM model is probably most advanced among the automatic test development methods. A test sequence is a sequence of input/output pairs that has been derived from the protocol specification, assuming a "Mealy" machine model. The goal is to test every transition of the underlying FSM at least once. The algorithms used to derive test sequences differ with respect to the assumptions made about the FSM serving as input, as well as for the length and error detection capabilities of the generated test sequences. A good overview is given in [Sidh89], where the most common methods are treated in detail (*transition tours, unique input/output sequences, distinguishing sequences, characterization sets*). Test generation environments providing tools for these methods are, for example, described in [Burg90] [Favr87] [Mizu88]. These tools accept an FSM format, Estelle, or SDL input.

One problem with all these FSM-based methods is that they do not take any data aspects into account. The question may be asked as to why each transition is tested only once and not several times with varying data parameters. Testing single transitions is not adequate, but possible event sequences and parameter dependencies have to be considered. A test derivation algorithm taking both control flow and data flow into account has been proposed in [Ural87] and realized in a corresponding tool. Protocol specifications in Estelle serve as input and are first transformed into a normal form. Then, the underlying state machine graph, as well as the data flow graph representing possible data flow between variables and interaction parameters are constructed. The test sequences derived from both graphs aim at covering all "definition-usage" pairs. This way, errors may be found that are related to variable usage without some defined value, or variable definitions that are not used by the protocol. Test case derivation from LOTOS specifications is still not as practical as the approaches described above. The theory developed in [Brin88b] has been implemented in the tools described in [Weze90] and [Alde90], but the *canonical testers* generated represent exhaustive sets of test cases that cannot be executed in practice. In [Tret90], the theory is restricted to *acceptance sets* and *reduced acceptance sets*, respectively. However, it is still unclear how these sets can be computed algorithmically [Tret90].

Another problem encountered for test case generation tools is the output format. Many tools use their own test description languages without any formal semantics. For international distribution and for gaining mutual recognition of test results by various parties, different test languages are not appropriate. Therefore, ISO is currently developing a standardized test notation called the *Tree and Tabular Combined Notation (TTCN)*. The definition of TTCN will become part of the ISO conformance testing standard [ISO9646].

Another complex activity in a heterogeneous world is **test suite execution** . Ideally, test suites should cover the full range of functions and capabilities of a protocol or service as defined in the specification. Since developers are free to make design choices and since they are often allowed to implement only subsets of the whole functional spectrum, test cases have to be selected from the test suite for each IUT. The selected tests have to be translated into an executable form. This translation may result in a program to be compiled, or yield statements in an executable test language. The tools described in [Wile89] and [Prob89] are able to process TTCN test descriptions in graphical (TTCN-GR) and in textual form (TTCN-MP) and to generate executable test cases. In [Baer89], the development of an interpreter for TTCN is presented, based on the informal TTCN semantics given in [ISO9646].

The ISO standard [ISO9646] also provides a framework for test case execution by defining several so-called *abstract test methods*, which are actually test configurations. The most important one is the *distributed test method*, depicted in Figure 65. The functionality of the tester is divided into an "upper tester" part, watching and controlling the upper service interface of the IUT, and a "lower tester" part, exercising control and observation of the IUT's lower service interface from a remote site. So-called *Points of Control and Observation (PCOs)* exist in the distributed test method between the IUT and the upper tester, and between the lower tester and an

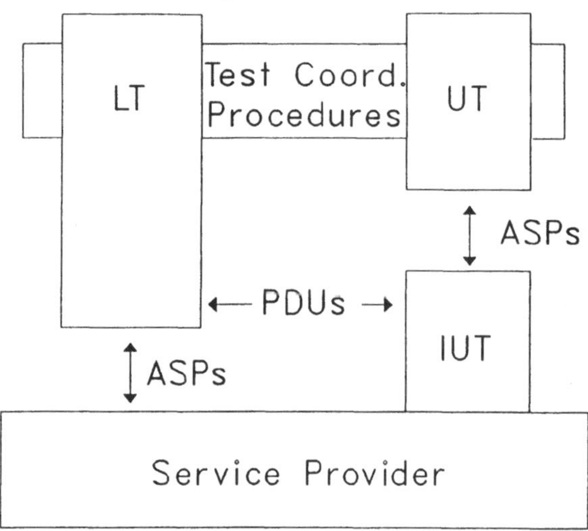

Figure 65. Distributed test configuration

underlying service provider. The lower tester resides in the system where the test is executed and is usually the major component of the whole test tool (*test driver*). Normally, the upper tester is a relatively small module which can be easily ported to different systems (*test responder*). The test system is connected to the IUT via the underlying service provider, which may involve a real network. Various forms of upper and lower testers are possible, however. A survey is provided in [Rayn87]. Test tools for the test methods introduced in [ISO9646] are described in [Mack88] [Chan90] [Stol90].

4.2.4 Integrated environments

The conclusion of this section on protocol engineering environments is that the integration of protocol engineering tools for all protocol development phases has not been properly achieved in most existing environments. This may be due to the fact that no FDT exists today which supports all the different phases of the development life-cycle equally well. On the other hand, a more general approach to the representation of the knowledge contained in protocol and service specifications is required to facilitate the integration of tools. The use of knowledge-based systems has been proposed, for example, in [Neum87] and [Shir88]. The rule-based nature of CRS specifications and the flexibility of X-ASN.1 data structures allow the realization of the ideas presented in [Neum87] in an advanced fashion. The architecture and realization of an integrated tools environment based on executable CRS and X-ASN.1 specifications is explained in the next chapter.

5 Protocol Engineering with Communicating Rule Systems

Several existing FDTs have been evaluated in chapter 2 to assess their suitability for specifying communication systems. Because of deficiencies in these FDTs, a rule-based technique for the specification of communication services and protocols has been proposed in chapter 3 (CRS, X-ASN.1). The advantages of CRS in relation to a number of general criteria for FDTs and the representation of architectural concepts of open systems have been explained. This chapter shows the relevance of using CRS and X-ASN.1 for building tools to automate protocol development. A protocol engineering approach is described that is based on the constructive nature of CRS specifications, allowing specifications to be executed according to the operational semantics defined in chapter 3.

A common characteristic of the protocol engineering environments described in chapter 4 is that the tools provided focus on particular phases in the development process instead of supporting the whole range of development activities. This is due to restrictions imposed by the techniques and by their corresponding execution mechanisms. As opposed to many other techniques, rule-based specifications may be executed with little overhead by employing well-established principles of rule-based programming. The rules represent the specification knowledge in a way that is independent of the intended application. As a consequence, the specification knowledge can be used by tools supporting the whole range of development activities. An implementation model and a runtime environment for CRS specifications is presented in section 5.1. This runtime environment forms the kernel of an integrated tools environment for protocol engineering, which is described in detail in section 5.2.

5.1 Executable CRS

As shown in the last chapter, tools supporting different protocol engineering activities can be realized on the basis of executable formal descriptions. The difficulty of executing formal descriptions depends on the level of abstraction of a particular FDT. Whereas a high level of abstraction is desired in order to facilitate implementation-independent specifications, it demands more effort to execute these specifications. In this section, the principles for executing CRS specifications are explained and some details for the realization of these principles are given.

5.1.1 Implementation Concepts

The basic idea behind the CRS implementation concepts is that of a *production rule-based system*, or short **production system** (see for example [Barr81] [Wins84] [Fros86]). In general, a production system consists of three major parts:

1. a *rule base*,
2. a *data base*, and
3. a *rule interpreter*.

The rule base is a set of *production rules* representing knowledge about the problem domain. Each production rule is of the form

if condition **then** action.

The data base is used to store information about the problem domain and its actual state provides the basis for evaluating the condition part of production rules. The actions of production rules may alter the state of the data base. The rule interpreter usually works in a cyclic fashion and repeats the following steps in each execution cycle: (1) find all rules in the rule base applicable to the present state of the data base (*matching*), (2) select one applicable rule (*conflict resolution*), (3) perform the actions of the selected rule (*action*). If the rule interpreter starts from an initial state and applies rules to work towards a goal state, it is called *forward-chaining*. If it tries to reason backwards from a goal state to an initial state, it is called *backward-chaining*. Since the rules in the rule base are used to infer new knowledge from existing knowledge, the rule interpreter in a production system is also called *inference machine* [Kric88].

CRS specifications can be directly implemented as production systems. The rules in rule system specifications define the rule base and the passive state objects constitute the data base. The mapping between specification and implementation is illustrated in Figure 66 for two rule systems (RS1, RS2) communicating over a common gate (Gate). Each rule system is executed by an associated CRS inference machine exercising control over the rule base. In order to facilitate communication between rule systems, each gate is handled by a CRS gate machine gathering event requests and event offers from the inference machines and looking for matching pairs.

For an adequate representation of objects in the data base an implementation concept for X-ASN.1 types and values has been developed based on so-called *structure trees* and *data trees* (see also [Mack88] [Sche09]). A detailed implementation concept for the efficient organization of the rule base and for the CRS inference and gate machines has been described in [Neum88]. These concepts are summarized in the following, before the CRS execution environment that has been realized based on this concepts is explained.

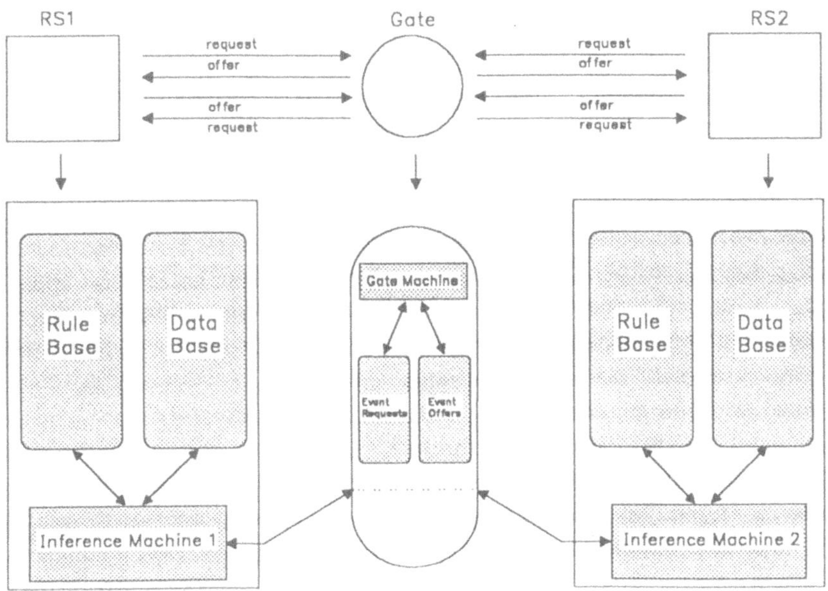

Figure 66. CRS implementation concepts

X-ASN.1 structure trees and data trees

In order to transform X-ASN.1 type definitions, instances, and operations into an executable representation, two approaches are possible:

- **translation** or
- **integration**.

In the **translation** approach, the X-ASN.1 type definitions are translated into corresponding types in a target language. Variable declarations are generated for each X-ASN.1 data instance and the X-ASN.1 operations are mapped onto operations, functions, or procedures in the target language. The ASN.1 compilers described in [Gora88a] and [Neuf90] are typical examples of tools realizing the translation approach, both using "C" as the target language. For some ASN.1 types, the translation into "C" is straightforward. For example, the simple types INTEGER, REAL, and ENUMERATED have direct counterparts in "C". Types constructed with the SEQUENCE and SET constructors are mapped onto structured types in "C". However, there are also ASN.1 simple types, as well as ASN.1 structured types, which cannot be translated directly because adequate constructs do not exist in "C" (or in similar programming languages).

One of the difficulties is given by the OPTIONAL construct in ASN.1. In the translation approach, a common solution is to generate an additional integer field for each optional component which holds information about the presence or absence of the optional component during runtime. Other difficulties arise out of the dynamic structures that can be defined with the SEQUENCE OF or SET OF constructors in ASN.1. Instances of such structures may shrink and grow as needed and their size requirements cannot be determined at compile time. The ASN.1 compiler described in [Neuf90] generates a special list structure (using "C" pointers to connect the components) for dynamic ASN.1 types constructed with SEQUENCE OF and SET OF. Similarly, the ASN.1 types NULL and ANY, as well as other predefined ASN.1 types for object identifiers or generalized time are translated into specialized "C" types. The ASN.1 OCTET STRING type is mapped onto a structured "C" type that contains additional fields for storing length and contents information.

Besides the difficulty of finding adequate representations for the ASN.1 types in the target language the main disadvantage of the translation approach is that the processing of data objects according to a protocol or service specification has to be implemented using the generated type definitions in the target language. Thus, checking the presence of optional fields, managing list structures, handling strings etc. is the responsibility of the implementor and is difficult to validate. A library of runtime routines can be provided to support developers, but some parts have to be generated for each ASN.1 type module (for example encoding/decoding routines) (see [Gora88a]).

In the **Integration** approach, X-ASN.1 is integrated into the target implementation language as a data definition language coexisting with the data type concept of the target language. Integration means that X-ASN.1 types can be defined and used just like types in the target language. Variables may be declared, values may be assigned, and operations may be invoked on data instances of X-ASN.1 types. In this way, an abstract syntax definition for a protocol or service can be used as the local syntax in an implementation. The representation of the X-ASN.1 data objects and the realization of the X-ASN.1 operations in the target language is hidden from the implementor. One way to support this approach is by employing a preprocessing tool that transforms all occurrences of X-ASN.1 variables and operations into the representation and realization in the target language, before the implementation code is compiled. Basic ideas for the representation of X-ASN.1 data objects in form of tree data structures are presented in the following. An X-ASN.1 preprocessor using this representation is described later in this section.

Representing X-ASN.1 type definitions and values in the form of tree data structures is appealing because of the dynamic properties of X-ASN.1 data objects. Tree structures are flexible and can be used to represent simple types (just one node), structured types (hierarchies or lists of nodes), and nested types (referenced subtrees). Nested types can be defined using other structured types as components, or using the same type as a component. In the latter case, recursive types are obtained and the tree structures representing them are actually graph structures.

The idea is illustrated in Figure 67: the X-ASN.1 type definitions are converted into an internal tree representation, called a **structure tree** (as for the type `Data` in the example). Structure trees are available for retrieving type information at runtime. The values of the different types are represented as so-called **data trees** corresponding to structure trees (example value representations are given on the right-hand side of Figure 67). Generic encoding/decoding routines can be designed to work for arbitrary X-ASN.1 data instances based on their underlying structure trees holding the type information.

Each component of an X-ASN.1 type definition is represented by a node in the corresponding structure tree. The nodes of structure trees are called **structure blocks** and contain type information as illustrated in Figure 68.

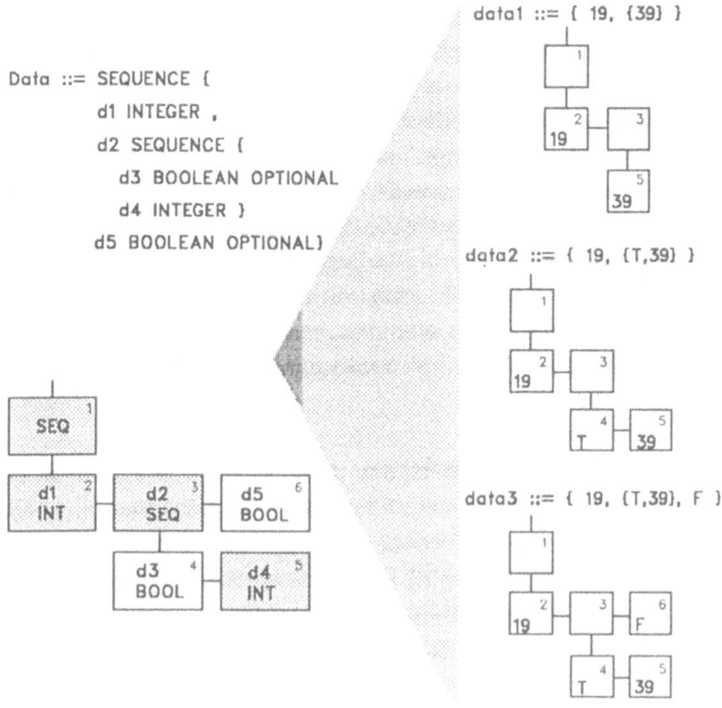

Figure 67. X-ASN.1 implementation concepts

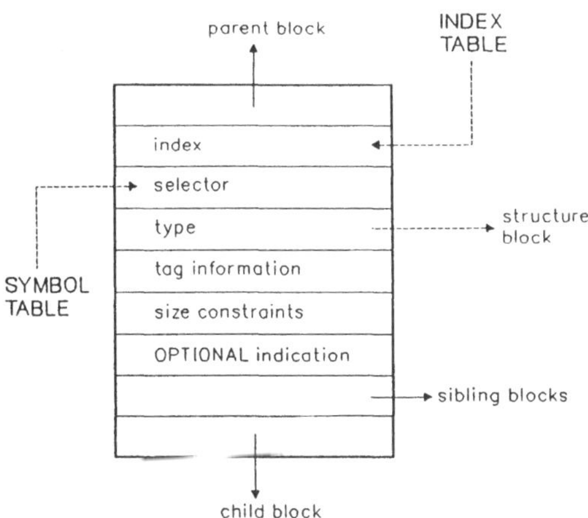

Figure 68. Structure blocks

Different fields in a structure block contain the component selector, the type of the component (a reference to another structure block), the tag class, number, and use of the IMPLICIT keyword, information about size constraints, and an indication whether the component is classified as optional. In addition, an index number is assigned to each structure block, so that the type representations are accessible via selectors from a symbol table, or via indexes from an index table. The remaining fields are pointers to parent, sibling, and child blocks in a structure tree.

An example of a complex structure tree is given in Figure 69, representing the CALL_REQUEST type defined in Figure 37 (see section 3.1). Structure blocks are drawn as boxes (if the structure block represents an optional component of the type the box is non-shaded). The structure block indices are written on top of the boxes. From this representation, it can be deduced that the CALL_REQUEST type is one alternative of the type X25_PDU constructed with the CHOICE constructor. The CALL_REQUEST type is tagged as context-specific, 11, implicit, and constructed as a SEQUENCE of fields as given in the child and sibling lists. Some of the fields are of simple types, others are of structured types again (the structured type FACILITIES is not shown in Figure 69).

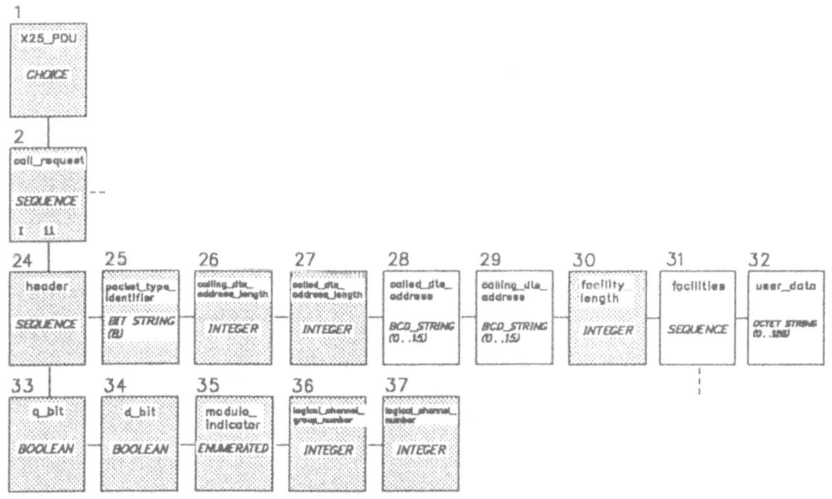

Figure 69. Structure tree for the X.25 CALL REQUEST packet type

Figure 70 illustrates the internal representation of a complete X-ASN.1 module. The various types defined in the module are represented as separate structure trees. Links between structure trees and even cycles may exist, representing nested type definitions and recursive types, respectively. A global symbol table contains all type identifiers and selector identifiers, whereas all indices of structure blocks are kept in a global index table. Application programs and encoding/decoding routines using the type information will start with the symbolic information, but may retrieve index numbers from it. Once the index numbers are available, access to the different nodes in the trees is much faster using the index table. The access mechanism is transparent to the application and mainly used by the internal functions working on the tree representations. All type information is loaded in an initialization phase and the required structure trees are configured.

Data trees refer back to their structure trees and are dynamically created and deleted during runtime. Arbitrarily many data tree representations may exist corresponding to a single structure tree, which is only represented once. The nodes of data trees are formed by **data blocks**, pointing to struc-

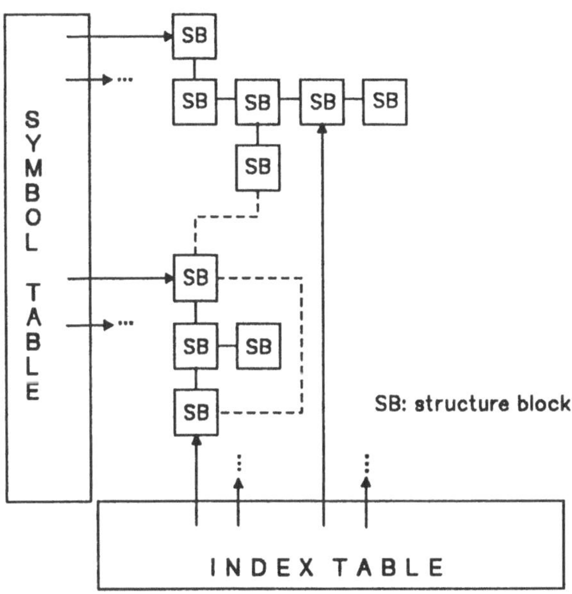

Figure 70. Structure trees

ture blocks (see Figure 71). For identifying the structure block a data block belongs to, the index number of the structure block is used. The value of the field associated with the data block is some concrete representation in the case of simple values, or a pointer to a subtree in the case of structured values. The remaining fields in data blocks are pointers to the parent, child, and sibling blocks.

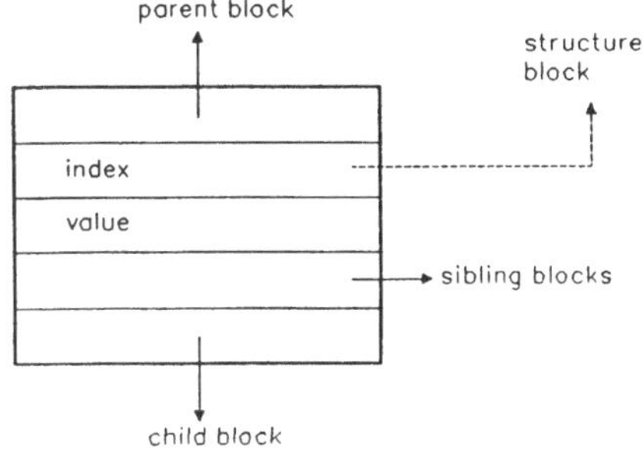

Figure 71. Data blocks

The representation of structured values requires several data blocks for the components to be linked according to the type structure (see Figure 73). A possible value of the CALL_REQUEST type (see Figure 69) is represented as a data tree in Figure 72. Inner nodes of the tree just carry the structure block index and pointers to the components, whereas actual value information is present in leaf nodes. In the particular instance depicted in Figure 72, the optional component facilities of the CALL_REQUEST type has been omitted. Data tree instances can be associated with an identifier and written to a file according to an external format. Using the identifiers, particular data trees can be loaded again at some later stage.

Since data trees point back to their corresponding structure tree, the tag information stored in structure blocks, together with the value information in data blocks, may serve as input to generic encoding/decoding routines. The encoding/decoding is guided by the type structure given in the structure tree. The routines convert from data trees as the local syntax to byte sequences in a transfer syntax (for example, defined by basic encoding rules) and vice versa. Since the type information is passed as a parameter

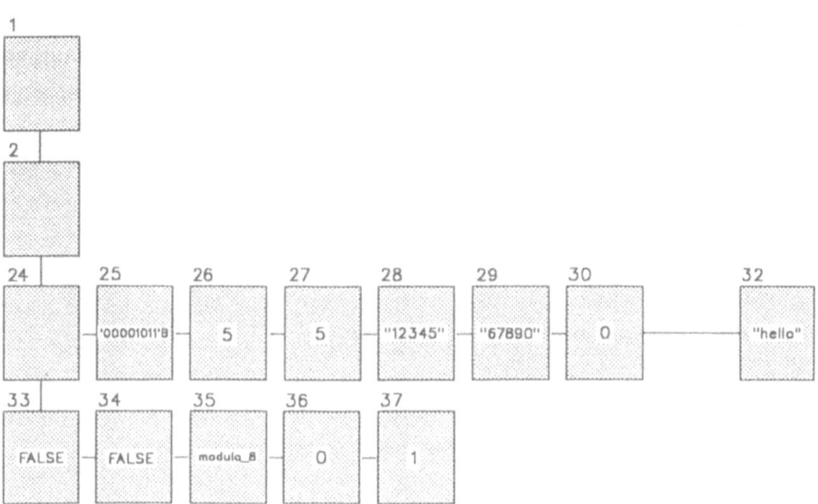

Figure 72. Data tree for an X.25 CALL REQUEST packet

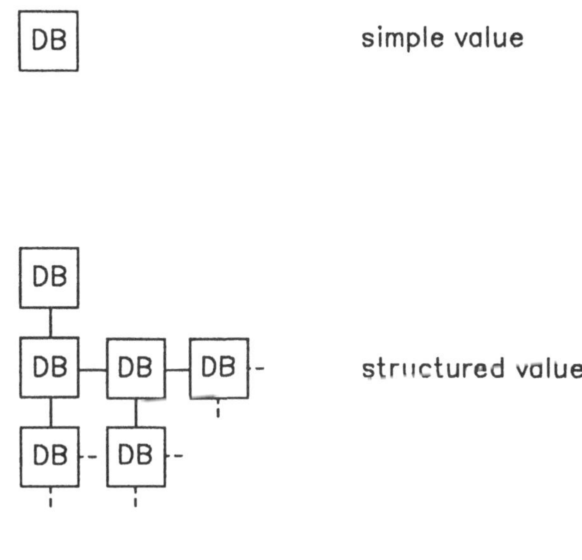

Figure 73. Data trees

to the routines, they can be designed in a generic fashion and need not be generated for each user-defined type.

CRS inference and gate machines

The CRS inference machines are based on the principle of forward-chaining. Given a current state of the rule system and a set of possible events, the set of rules that can be started or ended is determined (matching), one rule to be started or ended is selected (conflict resolution), and the corresponding actions are performed. In order to realize communication between inference machines, CRS gate machines keep track of all outstanding event requests and offers, and coordinate the attempts of the inference machines to communicate. Coordination is achieved by employing specialized protocols for the interaction between inference and gate machines at the implementation level. Two such protocols exist, serving for synchronous or asynchronous communication between inference machines. The CRS inference and gate machine concepts described in this section have been presented in [Neum88] and are suitable for implementing the CRS interleaving semantics defined in chapter 3. Only one rule may be started or ended at a time in each rule system and only a single event may occur in each step.

Each CRS inference machine organizes the rules of its associated rule system in different sets according to the execution state of the rule, as shown in Figure 74. Rules may be blocked, ready, or running, similar to processes managed by a dispatcher (see for example [Nehm85]). Besides the actual values of the state objects, the execution state of all rules and the values of the local objects of all running rules determine the state of a rule system. As long as no rule is being started or ended, all rules are in one of the three main rule sets:

Wstate which contains all rules that are not running and for which the start condition is false due to a non-qualifying state (*Wait for state*);

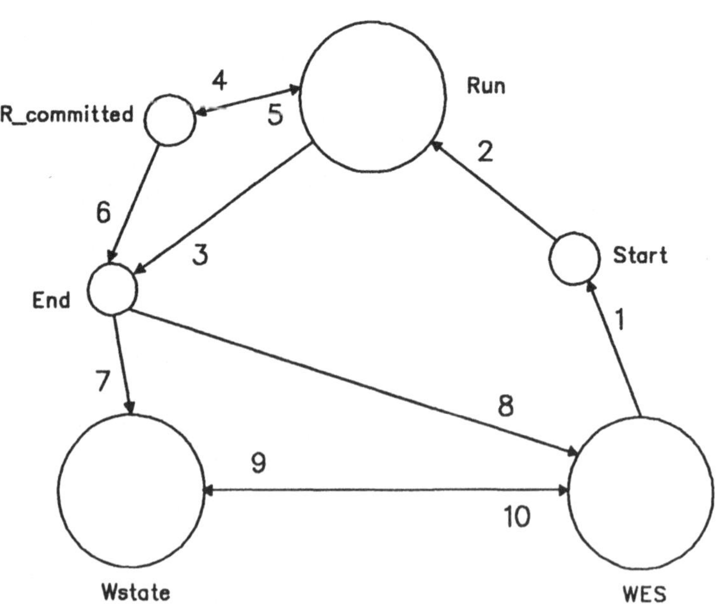

Figure 74. Rule sets of the CRS inference machine

WES which contains all rules that are not running and for which the start condition is true with respect to the current state, but the rules are requesting an outstanding event or are incompatible with some running rule (*Wait for Event or Synchronization*);

Run which contains all rules that have been started, but not yet ended (*Running*)[9].

The inference machine ensures that these rule sets are always maintained according to their meaning. Consequently, when a rule has been ended and the actions establishing the specified effects have been performed, rules may have to be transferred between the sets **Wstate** and **WES** (transitions 9 and 10 in Figure 74). Single element sets named **Start**, **R_committed**, and **End** serve as intermediate sets for transferring rules to and from the set **Run**. These intermediate sets are needed in case of synchronous communication as will be explained later. Startable rules are taken from the set **WES** and are transferred to the set **Start** (1). The acceptance of an event by the selected rule is indicated to the corresponding gate machine.

Before a rule is started, it remains in the **Start** set until the gate machine acknowledges the acceptance of the event after having informed the offering inference machine in case of a synchronous event (the details of the protocol are explained later). A rule is started by transferring it from the set **Start** to the set **Run** (2). Once being in the set **Run**, a rule is called "running".

A running rule with an asynchronous event (or without an event) in its effects is ended by transferring it directly to the set **End** (3), performing actions establishing the specified effects, and then passing it from the set **End** to the set **Wstate** (7) or **WES** (8), depending on the current state of the rule system. If a rule offers a synchronous event in its effects and a corresponding event request is given, the event parameters are passed to the gate and the rule is transferred to the set **R_committed** (4). If the event offer is accepted, the

[9] Remember that starting and ending a rule occur at two different points in time.

rule can be ended by transferring it to the set **End** (6). If the offer is refused, the rule is transferred back to the set **Run** (5). A new attempt to end the rule is made when a new event request arrives, or the values of the event parameters change.

If the three single element sets **Start**, **R_committed**, and **End** are empty, the rule system is in an idle state. Otherwise, at most one of these sets contains a rule and no actions can be performed for the rule system until an answer from the gate machine involved is received. The reception of gate machine messages may trigger a rule start from the set **Start**, a rule end from the set **End** via the set **R_committed**, or an unsuccessful termination from the set **R_committed**. Start condition checks and actions establishing effects can be executed atomically while the rule system state is fixed.

The set model is further detailed for the sets **WES** and **Run**. The **WES** set unfolds into three disjoint subsets as depicted in Figure 75. The **Ready** set contains the rules having no event request in their start condition (8b, 10a) and the rules coming from the set **Wsyn** (12). These rules may be immediately selected for a rule start (1a). Rules are transferred to the set **Wsyn** (8c, 10c, 11, 13) if they are not compatible with some running rule (see the definition of the compatibility relation in chapter 3). Rules waiting for an event are contained in the set **Wevent** (8a, 10b) and may be selected for a rule start if a matching offer is indicated by the corresponding gate machine (1b). These rules remain in the set **Wevent** if the event parameters are not acceptable (15). Rules in the set **Wevent** are transferred to the set **Wsyn** as soon as a rule is started with which they are not compatible (11). When the conflicting rule is ended, these rules reach the set **Ready** (12). If a matching event offer is still missing for these rules, they are again transferred to the set **Wevent** (14). If a rule being ended alters the rule systems' state objects in such a way that the start condition of a rule in the set **WES** is no longer satisfied, the rule is transferred from the set **WES** to the set **Wstate** (9a, 9b, 9c). Rules have to remain in the set **Wstate** if there is a matching event offer for them, but give a negative response to the gate machine during transition 17.

Similar to the set **WES**, the set **Run** is split into three sets: **R_fire**, **R_wfc**, and **R_wfc_p**, as shown in Figure 76. Rules with an asynchronous event offer (or

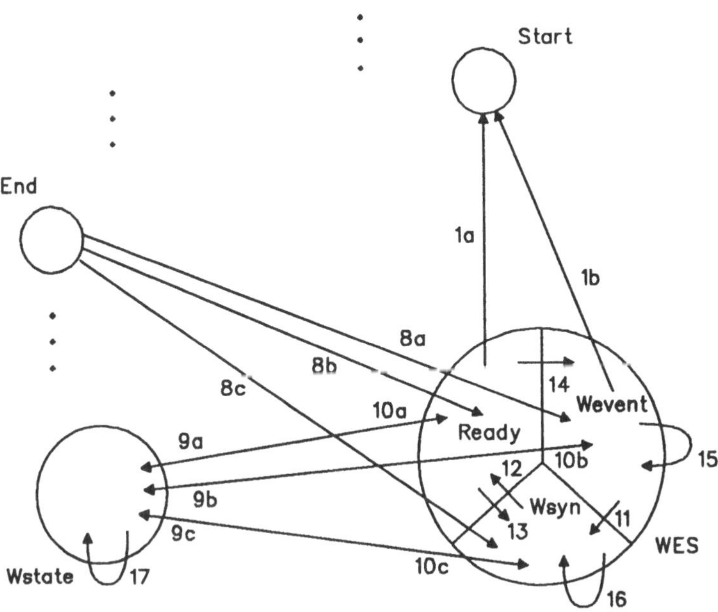

Figure 75. WES set of the CRS inference machine

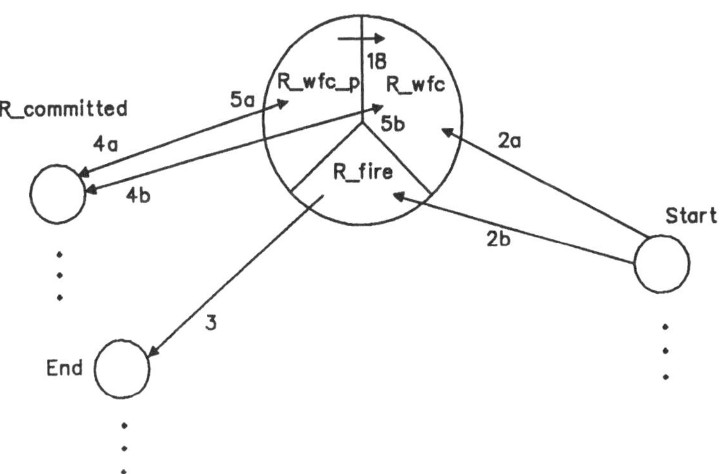

Figure 76. Run set of the CRS inference machine

without an event offer) are contained in the set **R_fire** (2b). Rules with a synchronous event in the effects are collected in the set **R_wfc** (2a). These rules have to be ended via the set **R_committed** as explained above (4b). If the attempt to end a rule is not successful because the offered parameters are wrong, the rule is put into the set **R_wfc_p** (5a). If other reasons prevented the successful termination, the rule is transferred to the set **R_wfc** (5b). From the set **R_wfc_p**, a rule may be selected again for termination (4a) if there is another matching request. Rules are transferred back to the set **R_wfc** (18) if the parameter values have been changed due to another rule end.

Organizing rules this way leads to increased efficiency of the CRS inference machines by means of a dedicated search mechanism. No pattern matching techniques are employed for searching rules to be started or ended. The different rule sets help restricting the number of rules to be examined at each step. The overhead of maintaining the rule sets can be kept small by using efficient implementation techniques for sets. Which rules are potentially enabled or disabled immediately after ending another rule can be determined by a static data flow analysis. This information can be utilized to increase the inference machine's intelligence [Neum88].

In order to implement synchronous and asynchronous communication on the CRS specification level a gate machine is associated with each gate instance on the implementation level. Gate machines store event requests and offers and check for matching events. As soon as a request and an offer that coincide in their event names are found, the gate machine coordinates the two inference machines concerned according to specialized protocols for synchronous and asynchronous communication. Each gate, and therefore each gate machine, can either support synchronous or asynchronous communication, not both.

The protocol automaton of the synchronous gate machine is depicted in Figure 77. The following states are possible:

idle no matching event requests and offers present, no new request or offer;

check_offers there is an actual request and a set of matching offers has to be checked;

wait_for_conf_offer one particular matching offer has been selected and the corresponding inference machine has to pass the actual parameters;

wait_for_conf_request actual event parameters have been passed and indicated to a requesting inference machine;

request_parms there is an actual offer and a set of matching requests;

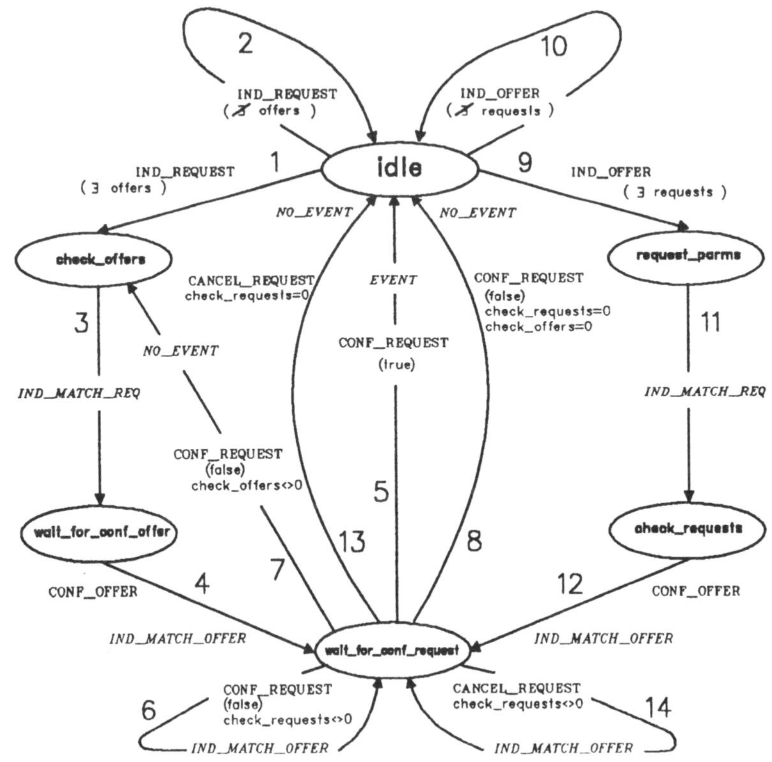

Figure 77. Protocol automaton of the synchronous CRS gate machine

check_requests	one particular matching request has been selected and the offering inference machine has to pass the actual parameters.

The possible incoming messages from inference machines are the following:

IND_REQUEST	a new event request arrives;
IND_OFFER	a new event offer arrives;
CONF_OFFER	an event offer is confirmed and actual parameters are passed;
CONF_REQUEST	an event request is confirmed (either positive or negative) with respect to the indicated actual parameters;
CANCEL_REQUEST	an event request is cancelled because the rule making the request is no longer startable.

The gate machine sends the following messages to coordinate inference machines:

IND_MATCH_REQUEST	a matching request is signalled to an offering inference machine;
IND_MATCH_OFFER	a matching offer is signalled to a requesting inference machine and actual parameters are indicated;
EVENT	the occurrence of the event is signalled to the participating inference machines;
NO_EVENT	an unsuccessful attempt to establish an event is signalled to the offering inference machine.

The protocol for a synchronous gate machine works as follows: in the state **idle**, arriving requests (message IND_REQUEST) are stored if no matching offers are found (transition 2 in Figure 77). If matching offers are present, a transition to the state **check_offers** is performed (1). Then, the message

IND_MATCH_REQ is issued and the state is changed to **wait_for_conf_offer** (3). After receiving the actual event parameters with the message CONF_OFFER, the presence of a matching offer is indicated to the requesting inference machine (message *IND_MATCH_OFFER*) and the state **wait_for_conf_request** is reached (4). However, this state may also be reached via the transitions 9, 11, and 12, which are performed if an IND_OFFER message arrives in the state **idle** and if matching requests are present. If no matching requests are present, the offer is stored and the protocol machine remains in the state **idle** (10). In the intermediate state **request_parms**, the message *IND_MATCH_REQ* is sent and the state **check_requests** is reached (11). After reception of the message CONF_OFFER, the message *IND_MATCH_OFFER* is sent and a transition to **wait_for_conf_request** is performed (12).

The protocol leaves open several possibilities to continue from the state **wait_for_conf_request**. A positive CONF_REQUEST message may arrive, causing the event to occur and the message *EVENT* to be sent to both inference machines (5). In case of a negative CONF_REQUEST message, one of the remaining matching offers is checked by performing a transition to the state **check_offers** (7). The message *NO_EVENT* is sent to the offering inference machine. Alternatively, there may be matching requests remaining for the offer received (6). The next request is selected and the message *IND_MATCH_OFFER* is sent. In this case, the state is not changed. If there are no matching offers or requests left, the message *NO_EVENT* is sent and the state is changed to **idle** again (8). A requesting inference machine which receives a matching offer indication, but finds that the corresponding rule is no longer startable in the current state of the rule system (the rule is contained in **Wstate**) returns a CANCEL_REQUEST message. Similar actions as for a negative CONF_REQUEST message are performed (13, 14). Additionally, the event request is cancelled at the gate.

The protocol for the asynchronous gate machine is a simplified version of the protocol just described. The protocol automaton is depicted in Figure 78. The message exchange between the gate machine and an offering inference machine is reduced to a single message (CONF_OFFER). With this message, an event offer is indicated and actual parameters are passed. The message IND_OFFER is not used. Also, the messages *NO_EVENT*

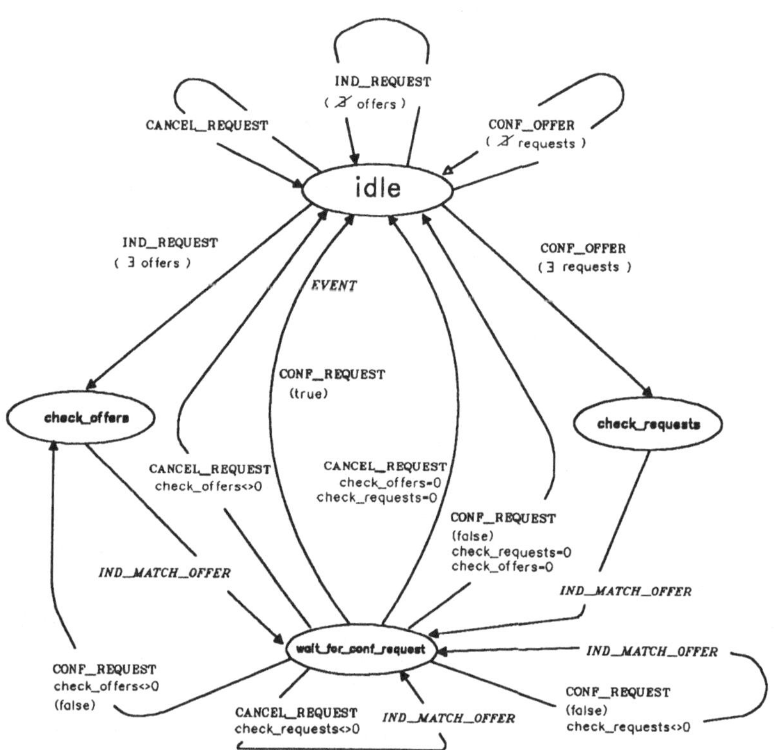

Figure 78. Protocol automaton of the asynchronous CRS gate machine

and *IND_MATCH_REQUEST* have no meaning in the case of asynchronous communication and they are not used.

These protocols implement synchronous as well as asynchronous communication correctly as provided by CRS and may serve for the local or distributed execution of communicating rule systems.

5.1.2 Runtime Environment

According to the implementation concepts explained in the last section, runtime support has to be provided for managing structure trees and data trees, as well as for the execution of rule systems and gates by inference and gate machines. The runtime environment provides the means for the simulated execution of CRS specifications in a distributed or local fashion. In a local realization, all rule system and gate instances defined in the specification are executed in a single address space and communication can be achieved using shared memory. Basic ideas of such a local realization are presented in the following. The different components of a local runtime environment have been implemented according to these ideas as described in [Sche89] [Schn87] [Schn89b].

Tree management

For executing CRS specifications which use X-ASN.1 data objects represented as structure trees and data trees, the runtime environment has to provide routines for

1. configuring and loading structure trees;
2. allocating and deallocating data trees;
3. navigating through structure trees and data trees;
4. assigning and copying data blocks.

In [Sche89], an implementation of these routines based on "C++" as the implementation language is described. By employing an object-oriented language, the tree management functions can be realized in an elegant way. Object classes defined for structure trees and data trees provide "constructor" and "destructor" functions that are inherited by specialized classes for individual X-ASN.1 types. The functions are used to allocate initial data trees when instances are created from these classes and free data trees when instances are deleted.

For each X-ASN.1 type constructor a class is defined which provides index and selection operations, respectively. Runtime checks are performed when navigating through structure trees and data trees. The value fields in data blocks can be changed by corresponding assignment functions. Copying whole data trees is supported, as well as structural assignment and comparison, by recursively applying the operations to subtrees. The tree management routines are called by the generic machines executing rule systems and gates.

Generic machines

CRS inference and gate machines can be provided in a generic fashion in a local runtime environment for CRS specifications by employing the concept of control blocks[10]. The execution state of rule system instances is represented by *rule system control blocks* and the execution state of gate instances is represented by *gate control blocks* (see [Schn87] [Schn89b]). The whole architecture of the local runtime environment for the simulated execution of CRS specifications is depicted in Figure 79.

A **generic inference machine** is able to execute arbitrary rule system instances by working on rule system control blocks. The functionality provided in the runtime environment comprises:

1. *matching*:
   - determination of startable rules;
   - determination of endable rules;
2. *conflict resolution*:
   - selection of rules to be started;
   - selection of rules to be ended;
3. *action*:
   - starting rules;
   - ending rules;
   - updating rule system control blocks.

The different functions work according to the rule set model described in the last section and implement the actions necessary for transferring rules between sets. Since many of these functions are independent from each other they can be realized as separate procedures. By combining these procedures differently inference strategies can be adapted to different needs. For establishing events functions of the generic gate machines are used.

---

[10] The idea of control blocks is similar to the *pseudo processors* proposed in [Nehm85] to implement process management in operating systems.

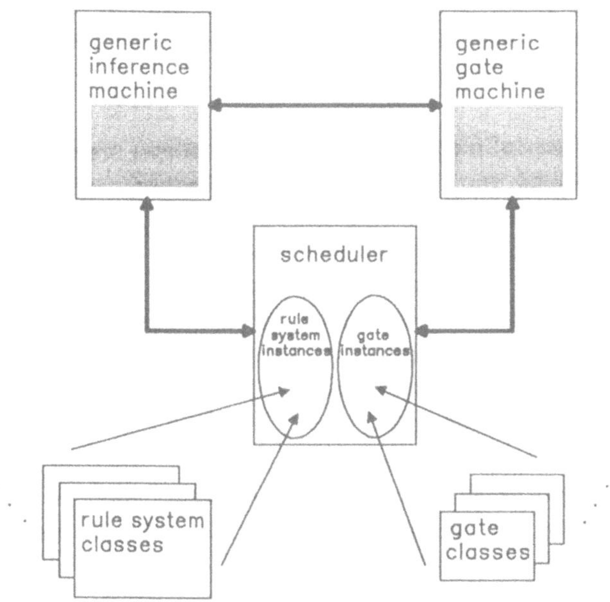

Figure 79. Architecture of the local CRS runtime environment

A **generic gate machine** is able to execute arbitrary gate instances by working on gate control blocks. However, separate generic gate machines are necessary for synchronous and asynchronous gates. They are implemented according to the protocols for synchronous and asynchronous communication between inference machines explained in the last section. The functionality provided in the runtime environment comprises:

1. handling event requests:
   - accepting new requests from inference machines;
   - indicating matching offers to requesting inference machines;
   - passing event parameters to requesting inference machines;
   - signalling event occurrences;
   - cancelling event requests;
2. handling event offers:
   - accepting new offers from inference machines;

- indicating matching requests to offering inference machines;
- requesting event parameters from offering inference machines;
- signalling event occurrences.

Rule system control blocks and gate control blocks are passed between the generic inference and gate machines by a **scheduler** in the runtime environment. The scheduler is responsible for managing rule system and gate control blocks after they have been created and for passing them to the generic machines according to the scheduling strategy used. The scheduling strategy should not be fixed for executing specifications, but it should be possible to adapt it to different needs. These needs may be determined by an application using the executable specification, or may be derived statically from the specification.

The information stored in rule system control blocks is summarized in Figure 80. Each rule set defined for the CRS inference machine is implemented as a list of *rule control blocks* and references to all rule sets are stored in the rule system control block. The name of the rule system instance is given in the rule system control block. Information about which rules are running is kept in the `Run` data structure. A boolean array indexed by rule numbers can be employed for this data structure. Such a representation is useful for checking the compatibility of startable rules with the running rules in an efficient way. Finally, the values of all state objects of the rule system are stored in the rule system control block.

As indicated above, important information about the execution state of a rule system is contained in *rule control blocks* (see Figure 81). If a rule is running, a reference to the values of the local objects is stored in the rule control block (`Lvars`). If a rule offers an event in the effects, a reference to the event parameter values is stored in `Parms`. A number of boolean `Flags` in rule control blocks are used to handle the communication with gate machines. Static information in rule control blocks comprises the name and an internal identification of the rule, as well as an identification of the rule system the rule belongs to. The `Syn` data structure represents the compatibility and priority relations for the rule. Similar to the `Run` data structure in rule system control blocks, boolean arrays indexed by rule identifications can be used. During runtime, compatibility and priority checks can be per-

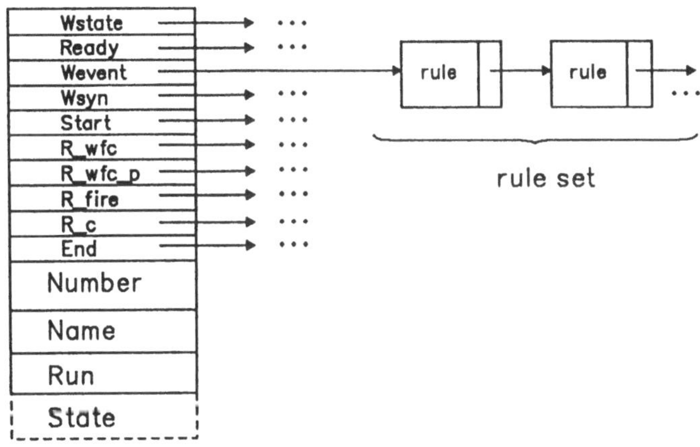

Figure 80. Rule system control block

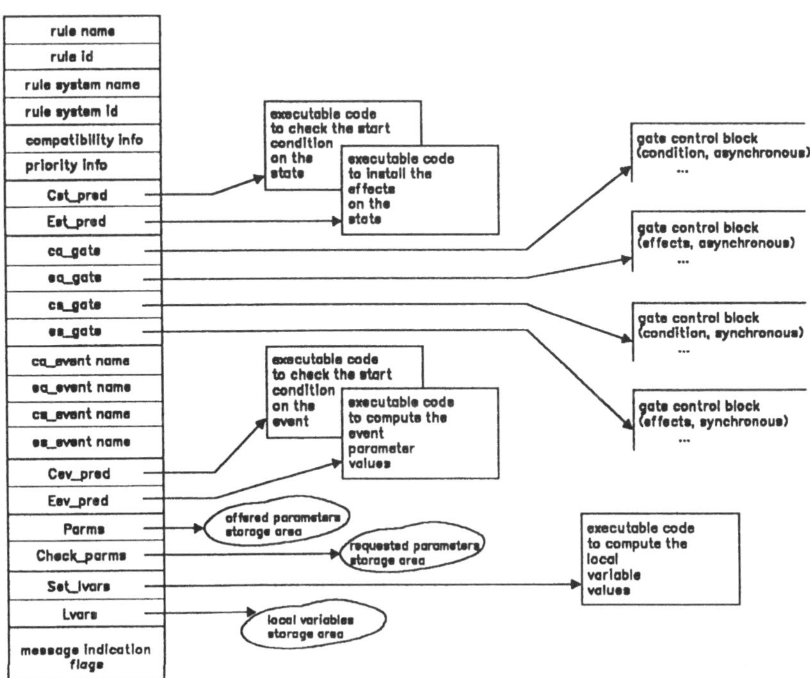

Figure 81. Rule control block

formed using simple logical operations on Run and Syn. If the rule requests an event in the start condition, or offers an event in the effects, the event types and references to gate control blocks are stored in the rule control block.

The $F_{cond\,(rs,\,r)}$ and $F_{effects\,(rs,\,r)}$ formulas of a rule are implemented by several functions and references to these functions are stored in the rule control block. According to the execution model used by the CRS inference machine (see Figure 74), the constraints on the rule system state and on events are separated. Each rule is implemented by five functions:

Cst_pred checks whether the state of the rule system qualifies for a rule start;

Cev_pred checks whether the offered event parameters meet the constraints of an event request;

Set_lvars determines settings of the local variables;

Est_pred performs actions establishing the effects on the state objects;

Eev_pred computes the event parameters to be offered.

The separation of start condition and effects parts into different functions is possible under the assumption made by the CRS implementation concepts that event requests and offers in rule start condition and effects parts are always $\wedge$-connected to the constraints on state objects. If no event is requested or offered, respectively, the corresponding functions are "empty" and simply return 'true'. The generic inference machine working on rule system control blocks invokes the functions in rule control blocks to check start conditions, save local variables, and realize effects during execution of a rule system. The results of these rule functions and the availability of gate messages determine whether rules can be transferred between the different sets.

The structure of gate control blocks is illustrated in Figure 82. The execution state of a gate comprises the state of the gate protocol and the event requests and offers available. Requests and offers stored in gate control blocks (Requests, Offers) are those that have been delivered to the gate by inference machines, but for which matching counterparts have not been found yet. Lists are used to represent sets of requests and offers. Some aux-

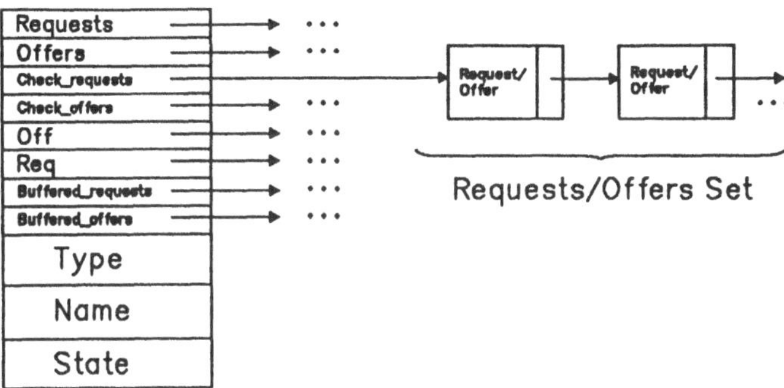

Figure 82. Gate control block

iliary sets are used for holding event requests to be checked against an offer and vice versa (Check_requests, Check_offers), an actual pair considered for a match (Req, Off), and requests/offers buffered until the gate machine returns to the idle state (Buffered_requests, Buffered_offers). The name of the gate and an indication of whether it is of synchronous or asynchronous type is static information in gate control blocks.

The elements of the request and offer sets are represented by *event control blocks* (see Figure 83). In each event control block, an identification of the event type is required (event name). Requested or offered event parameters can be retrieved via a reference to the rule making the request or offer, respectively.

Runtime support in form of tree management routines for X-ASN.1 objects, generic machines for rule systems and gates, and a scheduler controlling the activities forms the basis for the simulated execution of CRS specifications in a local environment. This runtime support is independent of the specification to be executed. Structure trees, data trees, control blocks, and

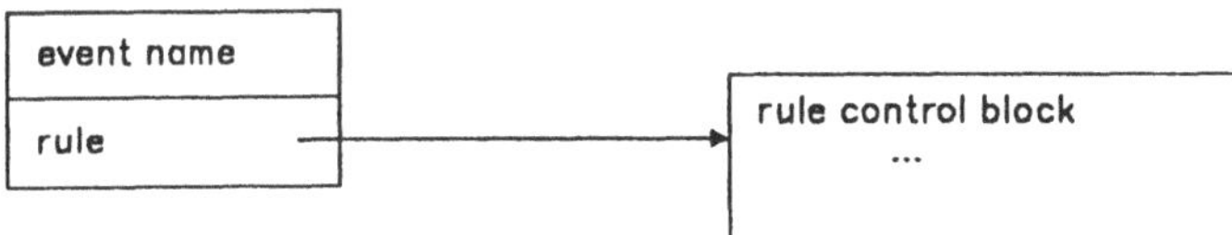

Figure 83. Event control block

the information contained in control block data structures have to be derived from individual specifications. This can be achieved by employing a compilation approach as described in the next section.

5.1.3 Compilation

The implementation of a compiler generating code for the local runtime environment from a given CRS specification (incorporating X-ASN.1) is described in [Clar89] and [Scho90]. After parsing and analyzing an input specification, an attributed syntax tree and a global symbol table are built. The code is generated from this internal representation and comprises type definitions for the state objects of rule systems and ADTs, local objects, and event types, as well as five functions for each rule and one function for each ADT operation. Initialization procedures are generated for creating control blocks and for storing the initial values of the state objects.

The approach taken is to generate separate modules in the target language for each gate, rule system, and ADT specification, and one additional module for creating the specified configuration of rule system and gate instances. The target language used is an extended form of "C++" code. The extensions are mainly concerned with X-ASN.1 types and variables, following the X-ASN.1 integration approach as described in section 5.1.1. Some concrete examples are given later. Each module consists of a **header file** and a **source file**. The header file contains statements defining the interface of the module and the types and procedures exported to other modules. The source file contains the implementation of these types and procedures.

The different steps in the compilation process are illustrated in Figure 84. A CRS specification is parsed and analyzed by the CRS compiler front-end and the internal representation is built. The X-ASN.1 type definitions used in the specification are extracted and serve as input to an X-ASN.1 compiler generating structure tree representations according to the concepts presented in section 5.1.1. These structure trees are used by the CRS compiler back-end generating the X-ASN.1 extended "C++" code for the specification. The extensions are removed by the X-ASN.1 preprocessor and the code is translated to plain "C" code by a "C++" preprocessor. The final "C"

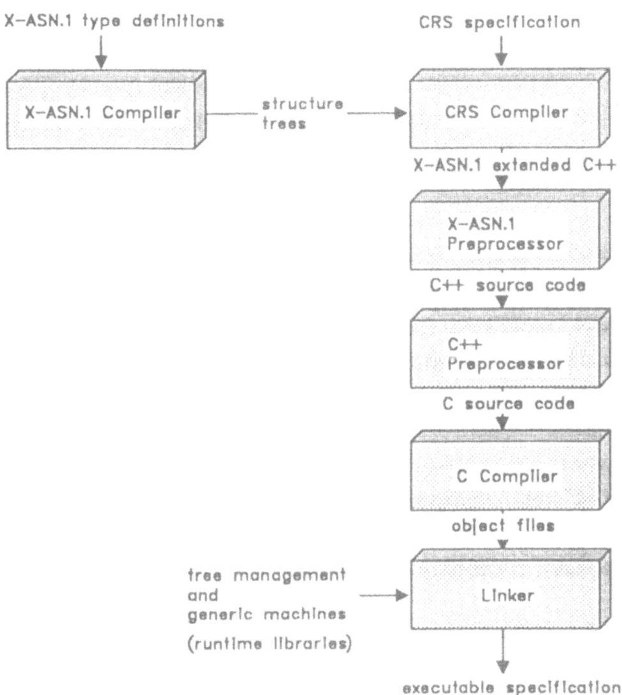

Figure 84. Compiling CRS specifications into executable form

sources are compiled and linked with runtime libraries providing the functions described in section 5.1.2 to form an executable specification.

In the following, the principles employed for the compilation of the different parts in a CRS specification are explained in more detail.

X-ASN.1 preprocessing

In section 5.1.1, the integration of X-ASN.1 into a target implementation language and the representation of types and values as structure trees and data trees has been described. One way to realize this approach is by employing a preprocessing tool that transforms all occurrences of X-ASN.1 variables and expressions into the target language representation, before compiling the sources into object code. The tree representation is hidden to programmers or code generator tools, respectively, using X-ASN.1 as the data definition language. The X-ASN.1 access operations and variable manipulations are transformed into appropriate tree management function calls by the preprocessor. The tree management functions are provided in a runtime library. The mixed usage of X-ASN.1 and data definition concepts of the target language is possible. Such a tool has been implemented as described in [Sche89].

An example of X-ASN.1 extended "C++" code as generated by the CRS compiler is shown in Figure 85. Two types are defined in this example, one using X-ASN.1 (ASN_MY_TYPE) and the other using "C++" constructs (C_MY_TYPE). The first component of C_MY_TYPE is of the type ASN_MY_TYPE and can be accessed using the selector identifier asn_my_sel. The variable mixed_var of the type C_MY_TYPE is declared and used in an assignment statement, where the value of the X-ASN.1 component is changed. The left-hand side of this assignment statement consists of a mixed access expression, starting with a "C++" selection operation and continuing with X-ASN.1 access operations following the sensitive selector identifier

```
ASN_MY_TYPE ::= SEQUENCE {
                    text   SEQUENCE OF IA5STRING,
                    flag   BOOLEAN }

typedef  struct { ASN_MY_TYPE   asn_my_sel;
                  int           i;
                } C_MY_TYPE;

C_MY_TYPE   * mixed_var;


    mixed_var -> asn_my_sel . text [1] = "hello";

//  <--------------------->
//       "C" access
//               <-------->
//               sensitive
//               selector
//               <------------------->
//                   X-ASN.1 access
//                             <------>
//                              X-ASN.1
//                              component

    LeftHandAccess ( mixed_var->asn_my_sel, 37, 1, 0 ) = "hello";
```

Figure 85. Example of X-ASN.1 extended "C++" code

asn_my_sel. The various parts of this access expression are marked in Figure 85. The function call generated for this access expression by the X-ASN.1 preprocessor is given at the bottom of Figure 85. The first argument of the function LeftHandAccess is a "C++" expression referring to an X-ASN.1 variable. The other arguments are block indices of the underlying tree representation (see section 5.1.1) and an indication of the access mode.

The preprocessor works in two phases, called *marking phase* and *transformation phase*. In the marking phase, the X-ASN.1 types and expressions are analyzed and type compatibility is checked. Type, variable, and selector identifiers are stored in a symbol table. In the transformation phase, all X-ASN.1 related statements are transformed into their "C++" counterparts.

The implementation of the runtime support routines for the tree management (see section 5.1.2) is also described in [Sche89]. The class hierarchy is outlined in Figure 86. The most general class is ASN_EXPR, fixing the repre-

```
//  ASN_EXPR <----- ASN_EXPR_INT   <----- ASN_VAR_INT
//           <----- ASN_EXPR_REAL  <----- ASN_VAR_REAL
//           <----- ...            <----- ...
//
//           <----- ASN_EXPR_OTHER <----- ASN_VAR_OTHER

class ASN_EXPR {
  DT   data_tree;
  DB   data_block;
};

class ASN_EXPR_INT : public ASN_EXPR {
 public:
  void   operator =  (ASN_EXPR& source);                        // assignments
  void   operator =  (int i);
  int    operator == (ASN_EXPR& second);                        // comparison
  int    operator int();                                        // conversion
}

class ASN_VAR_INT : public ASN_EXPR_INT {
 public:
   ASN_VAR_INT (int st_index, int sb_index, char * id, bool empty)// constructor
   ~ASN_VAR_INT ()                                              // destructor
}
 ...
```

Figure 86. "C++" classes for X-ASN.1 expressions and variables

sentation of an X-ASN.1 object by a data tree and a data block. Objects of the class `ASN_EXPR_INT` derived from `ASN_EXPR` inherit this representation. Operations for assignment and comparison, as well as for conversion to the "C++" integer type are defined. In this way, all arithmetic and relational operations are mapped onto their "C++" equivalents. Finally, constructor and destructor functions for X-ASN.1 integers exist for the class `ASN_VAR_INT` derived from `ASN_EXPR_INT`. All other X-ASN.1 types are handled similarly. For the constructed types structural assignment and comparison is supported.

Compilation of gates

The compilation process for gates results in a header file and a source file for each gate specification. The attributed syntax tree built during the parsing phase serves as the basis for generating these files. As an example, the header file and source file for the gate `Network_service_access_point` (see Figure 54 in chapter 3) as generated by the CRS compiler are outlined in Figure 87 and Figure 88.

The header file contains type definitions for each event type defined for the gate (see part `II` in Figure 87). The X-ASN.1 type modules which are required for defining these types are included by corresponding X-ASN.1 preprocessor directives (`I`). All identifiers generated by the compiler (type identifiers in this example) are constructed based on the identifiers used in the specification text. Besides the type definitions, the interface description of a 'copy'-function is generated for each event type defined in a gate specification (`III`). The 'copy'-functions for event types are called by the generic runtime routines in order to copy event parameter values between memory areas when requesting and offering events at runtime. The structure of these memory areas does not have to be known by the generic machines and the implementation of the 'copy'-functions for event types can be hidden in the gate source file. Another function declaration is contained in the gate header file for the 'create'-function, serving for the creation of gate instances (`IV`). The 'create'-function is called whenever a new instance of a gate specification has to be allocated and initialized. The implementation of this function is given in the gate source file.

At the beginning of the gate source file, the gate header file is included to make all the type definitions available (see part I in Figure 88). Then, the code generated for the event parameter 'copy'-functions follows (II). The N_CONNECT_req_ev_copy-function is shown as an example in Figure 88. Two generic pointers (from, to) are given as parameters. These pointers refer to objects whose sizes and structure are unknown outside the function. Within the function, the addresses passed in the parameters are assigned to local pointers (pev, from_casted) of appropriate types (N_CONNECT_req_ev).

The 'copy'-functions serve several purposes in the runtime environment. Depending on the parameter values given, they return size information, allocate memory, copy values, or free memory. After the code of the 'copy'-functions, the code generated for the 'create'-function is given in the gate source file (III). The name of the new gate instance is passed as a parameter. A new gate control block is allocated (function NEW_SGATE provided by the runtime environment) and the initial values are supplied.

```
//-- I -------------------------------------- X-ASN.1 preprocessor directives --
$add_asn_type_module  ASN1_STANDARD_TYPES
$add_asn_type_module  NetworkService
//------------------------------------------------------------------------ I --

//-- II ------------------------------------------------ event type definitions --
typedef struct {
    NetworkService.N_CONNECT_request *asn_n_connect_req;
  } N_CONNECT_req_ev;

typedef N_CONNECT_req_ev *PN_CONNECT_req_ev;
 ...
//-----------------------------------------------------------------------II --

//-- III -------------------------- external declarations for COPY-functions --
extern int N_CONNECT_req_ev_copy (Gen_pointer from, Gen_pointer *to);
 ...
//----------------------------------------------------------------------- III --

//-- IV --------------------------- external declaration for CREATE-function --
extern PSynGateset CR_Network_service_access_point (char *name);
 ...
//------------------------------------------------------------------------ IV --
```

Figure 87. Example gate header file

```
//-- I -------------------------------------------- include gate header file --
#include "n_sap.h"
//--------------------------------------------------------------------- I --

//-- II --------------------------------------------------- COPY-functions --
int N_CONNECT_req_ev_copy (Gen_pointer from, Gen_pointer *to)
{
   N_CONNECT_req_ev *pev;
   N_CONNECT_req_ev *from_casted;

   if ((from == NULL) && (to == NULL)) return (sizeof *pev);

   if (to== NULL) {
      from_casted = (N_CONNECT_req_ev *) from;
      delete from_casted->asn_n_connect_req;
      free (from_casted);
      return (0);
   } /* endif */

   pev = (N_CONNECT_req_ev *) malloc (sizeof *pev);

   if (pev != NULL) {
      new_asn_var pev->asn_n_connect_req;
      if (from!= NULL) {
         from_casted = (N_CONNECT_req_ev *) from;
         *pev->asn_n_connect_req = *from_casted->asn_n_connect_req;
      } /* endif */

      *to = (Gen_pointer) pev;
      return (0);
   } else {
      return (1);
   }
}
 ...
//--------------------------------------------------------------------- II --

//-- III -------------------------------------------------- CREATE-function --
PSynGateset CR_Network_service_access_point (char *name)
{
   PSynGateset p;
   ...
   p        = NEW_SGATE (name);
   p->state = idle;
   ...
   return(p);
}
//--------------------------------------------------------------------- III --
```

Figure 88. Example gate source file

Compilation of rule systems

The compilation process for rule systems results in a header file and a source file for each rule system specification. The various parts of the specification (STATE, DECLARATIONS, DEFINITIONS etc.) are parsed and an attributed syntax tree is built (see [Clar89]). The names of parameters, objects, rules etc. are stored in a symbol table.

The code generation starts from the symbol table and an attributed syntax tree corresponding to a specification which is **expanded**, **resolved**, and **normalized**. This means, auxiliary predicates, contexts, and views are removed in a number of passes (see [Mill90]), resulting in a flat rule system (see chapter 3). Additionally, all formulas used in the specification are transformed into *prenex-disjunctive normal form* by employing a corresponding algorithm (see for example [Mann74]).

An example of a header file generated from a rule system specification is given in Figure 89 for the rule system X_25_DTE specified in chapter 3 (see Figure 48). All X-ASN.1 type modules used are included by X-ASN.1 preprocessor directives (see part I in Figure 89). The state objects of the rule system are summarized in a corresponding type definition and another type is defined for all local objects (II, III). An external declaration for the 'create'-function of rule system instances is given (IV).

The types defined in the header file are included in the source file generated from a rule system specification (see Figure 90, part I). Other header files are included for all gates used by the rule system. Similar to the event parameter 'copy'-functions given in the source file for gates, 'copy'-functions for the state objects and local objects of a rule system are generated (II). As explained in section 5.1.2, five functions are generated for each rule in the rule system specification, serving to check start conditions and to realize effects (III). How these functions are obtained is described later. An example is given in Figure 93 for the rule n_con_req (see Figure 50).

Finally, the implementation of the 'create'-function is given in the source file generated from a rule system specification (IV). The formal parameter list of this function corresponds to the formal parameter list of the rule system

```
//-- I ------------------------------------- X-ASN.1 preprocessor directives --
$add_asn_type_module  ASN1_STANDARD_TYPES
$add_asn_type_module  X_25_Packets
$add_asn_type_module  NetworkService
$add_asn_type_module  X_25_DTE
//------------------------------------------------------------------------ I --

//-- II ---------------------------------- state variable type definitions --
typedef struct {
   R_state_type *asn_r_state;
   ...
   } X_25_DTE_state;

typedef X_25_DTE_state *PX_25_DTE_state;
//----------------------------------------------------------------------- II --

//-- III ---------------------------------- local variable type definitions --
typedef struct {
   N_CONNECT_request     * asn_N_CR_SP;
   CALL_REQUEST          * asn_CR_PDU;
   ...
   } X_25_DTE_decl;

typedef X_25_DTE_decl *PX_25_DTE_decl;
//---------------------------------------------------------------------- III --

//-- IV --------------------------- external declaration for CREATE-function --
extern PRsset CR_X_25_DTE (char *name, PSynGaterepr Nsap, PSynGaterepr Packets,
                           Logical_channel_number LC);
//----------------------------------------------------------------------- IV --
```

Figure 89 Example rule system header file

specification. The following actions are performed by the function: a rule system control block is allocated (function NEW_RS provided by the runtime environment), memory for the state objects is allocated (using the 'copy'-function for state objects), and the initial values are supplied according to the **INITIALLY**-clauses given in the specification.

After the rule system control block has been allocated and initialized, rule control blocks are allocated for each rule (function NEW_RULE provided by the runtime environment). The rule-specific data is passed in a number of parameters. Then, the new rule control block is inserted into the **Wstate** set of the rule system control block.

```
//-- I --------------------------- include gate and rule system header files --
#include "n_sap.h"
#include "x_25_acc.h"
#include "x_25_dte.h"
//------------------------------------------------------------------------ I --

//-- II ---------------------------------------------------- COPY-functions --
int X_25_DTE_STATE_COPY (PRsrepr prsold, PRsrepr * prsnew)
{ ... }
int X_25_DTE_DECL_COPY (Gen_pointer from, Gen_pointer *to)
{ ... }
//----------------------------------------------------------------------- II --

//-- III ------------------------------------------- rule body procedures --
// see separate figure
//------------------------------------------------------------         III --

//-- IV ---------------------------------------------------- CREATE-function --
PRsset CR_X_25_DTE (char *name, PSynGaterepr Nsap, PSynGaterepr Packets,
                    Logical_channel_number LC)
{
  ...
  p  = NEW_RS (name, nr, X_25_DTE_STATE_COPY);
  rc = X_25_DTE_STATE_COPY(NULL, &(p->rspart));
  *state_area->asn_r_state = r1;
  ...
  help = NEW_RULE("n_con_req",
              0,                    /* number of rule (internal)             */
              p->rspart,            /* pointer to surrounding rule system    */
              help_vector,          /* internal synchronization information */
              n_con_req_CST,        /* Cst_pred function                     */
              n_con_req_EST,        /* Est_pred function                     */
              NULL,                 /* asynchronous start condition gate     */
              NULL,                 /* asynchronous effects gate             */
              Nsap,                 /* synchronous start condition gate      */
              Packets,              /* synchronous effects gate              */
              "",                   /* asynchronous start condition event    */
              "",                   /* asynchronous effects event            */
              "N_CONNECT_req",      /* synchronous start condition event     */
              "Call_request",       /* synchronous effects event             */
              n_con_req_CEV,        /* Cev_pred function                     */
              n_con_req_EEV,        /* Eev_pred function                     */
              Call_request_ev_copy,/* Parameter COPY-function               */
              X_25_DTE_DECL_COPY,   /* local variables area COPY-function    */
              n_con_req_SLV);       /* Slv_pred function                     */
   INS_RULE(&(p->rspart->Wstate),help);
   ...
//---------------------------------------------------------------------- IV --
```

Figure 90. Example rule system source file

Compilation of rules

Most of the complexity encountered for the compilation of CRS specifications is due to the first order predicate logic (FOPL) used for specifying the start conditions and the effects of rules. According to the implementation concepts for the CRS inference machine explained in section 5.1.1, code has to be generated to evaluate the start condition with respect to state and event constraints, to determine appropriate values for local objects, to perform actions on the state, and to offer events for each rule. Appropriate restrictions on the form of predicates used in rule specifications had to be found in order to make an automatic translation feasible.

As already mentioned above, the start condition and effects formulas of rule specifications ($F_{cond\,(rs,\,r)}$ and $F_{effects\,(rs,\,r)}$, see section 3.3) are transformed into *prenex-disjunctive normal form (DNF)*, before the code generation phase starts. Quantors are not supported. In general, the formulas are then of the form

$$C_1 \lor C_2 \lor \ldots \lor C_n \quad .$$

Each C_i (i=1..n) is a conjunctive sub-formula of the form

$$L_1 \land L_2 \land \ldots \land L_m$$

and represents one alternative for satisfying the whole formula. For a C_i to be satisfied each L_j (j=1..m) must be satisfied. The L_j are called *literals* and are of the form

$$A \quad \text{or} \quad \neg A \ ,$$

where A is an *atomic formula.* Atomic formulas in FOPL (with equality) are of the form

$$p(t_1, \ldots, t_k) \quad \text{or} \quad t_1 = t_2 \quad ,$$

where $t_1, \ldots, t_k$ are terms. Terms may be just variables, or some function constant (for example, +, /, etc.) applied to terms. The symbol p represents a predicate constant (for example, $\leq$, $>$, etc.).

The DNF form of formulas is more suitable than a conjunctive normal form because the implementation concepts presented in section 5.1.1 are based on the assumption that event requests and offers are always $\wedge$-connected to the rest of the predicate and that state and event requirements are checked separately. Code can be generated from the DNF for each alternative $C_1, ..., C_n$, where an interaction expression, if present, has to be part of every alternative and is always $\wedge$-connected to state requirements and constraints on event parameters. Most preferably, start conditions and effects formulas consist of just one alternative C_1 and several alternatives are expressed using several rules, instead of $\vee$-connected sub-formulas.

The semantics of formulas in FOPL is given by means of **interpretations** as already explained in section 2. In the following, a notation similar to the one in [Mann74] is used. An interpretation $\boldsymbol{I} = (D, I_c, I_v)$ consists of a domain D and mappings for the constants and free variables in a formula, where

I_c assigns

- elements of D to individual constants,
- value 'true' or 'false' to propositional constants,
- a function over D to function constants,
- a predicate over D to predicate constants.

I_v assigns

- elements of D to individual variables,
- value 'true' or 'false' to propositional variables.

An interpreted formula (association of interpretation and formula) always has a truth value *true* or *false*. An interpretation yielding *true* is called a **model** of the formula.

The domain D of possible interpretations for $F_{\mathit{cond}\,(rs,\,r)}$ and $F_{\mathit{effects}\,(rs,\,r)}$ is given by the cartesian product over the X-ASN.1 types of the variables used in the formulas. The constants, functions, and predicates are mapped onto the corresponding X-ASN.1 constants and operations. For the variable assignment three sorts of variables are distinguished:

- state variables,
- local variables, and
- auxiliary variables.

The state variables are declared in the STATE-part of a rule system specification. Both local variables and auxiliary variables are declared in the DECLARATIONS-part of a rule system specification. They are called "declaration variables" in the following. Local variables are those declaration variables that are used both in the $F_{cond\,(rs,\,r)}$ and $F_{effects\,(rs,\,r)}$ formula. Auxiliary variables are either used in $F_{cond\,(rs,\,r)}$ or $F_{effects\,(rs,\,r)}$, but not in both formulas.

The code generated for the five functions implementing a rule serves to check models of $F_{cond\,(rs,\,r)}$ and $F_{effects\,(rs,\,r)}$, respectively. Depending on the particular function, possible models are partly defined by the state of the rule system, the offered event parameters, or the values of local variables of running rules. The missing parts of a model have to be determined by a static analysis of the formula. Appropriate values obtained from such an analysis are propagated into local variables, the new state of the rule system, or event parameters being offered. The approach taken in [Scho90] to realize the generation of code for checking (respectively constructing) models of rule specifications is further detailed in the rest of this section.

Starting from $F_{cond\,(rs,\,r)}$ and $F_{effects\,(rs,\,r)}$ given in DNF, variable occurrences in literals (L_j) are separated into **defined occurrences** and **used occurrences.** In defined occurrences, possible values of a variable are restricted by a predicate. For example,

```
x < 5
```

is a literal, where the value of the variable x is restricted to the set {4, 3, 2, ...} by the < predicate. In general, several defined occurrences of the same variable may be given in a formula. For example, in the formula

```
x < 5  ∧  x > 3
```

the combination of two defined occurrences restricts the value of x to "4". However, the computation of combined constraints is, in general, a very complex task. The tool described in [Scho90] recognizes only defined occurrences of the form $t_1 = t_2$, where either t_1 or t_2 is a variable, or both are variables. For these forms no symbolic computation is required.

In a formula like

```
x < 5  ∧  x > 3  ∧  y = x
```

the first two occurrences of x can be regarded as defined occurrences and the third occurrence is a used occurrence: the variable x (being already defined) is used to define the value of the variable y.

Classifying the variable occurrences in a formula this way does not necessarily deliver a unique result. For example, in the formula

```
x < 5  ∧  x > 3  ∧  y = x  ∧  y = 4
```

the third literal can also be regarded as a defined occurrence of x and a used occurrence of y. The first two literals would then be regarded as used occurrences of x. In any case, the static analysis ends successful and a model of the formula is determined.

In the literal

```
x + 1 = 2
```

the occurrence of x may also be seen as a defined occurrence. Appropriate values for x can be found by solving the equation. However, combining several constraints on a number of variables this way leads to equation systems, which need not be linear. Manipulations of this kind were considered the cause of too much overhead in the tool described in [Scho90]. The experience gained during several specification case studies with CRS in the field of protocols showed that the need to solve equations in order to identify defined occurrences of variables rarely arises in practice.

Before the code is generated, the $F_{\textit{cond}\,(rs,\,r)}$ and $F_{\textit{effects}\,(rs,\,r)}$ formulas of a rule specification are analyzed to find the variable dependencies. During several passes over the attributed syntax tree of a formula, a *dependency graph* is built for each conjunction.

This is illustrated by the dependency graph of the rule n_con_req shown in Figure 91 (see the specification given in Figure 50). In the general case (a

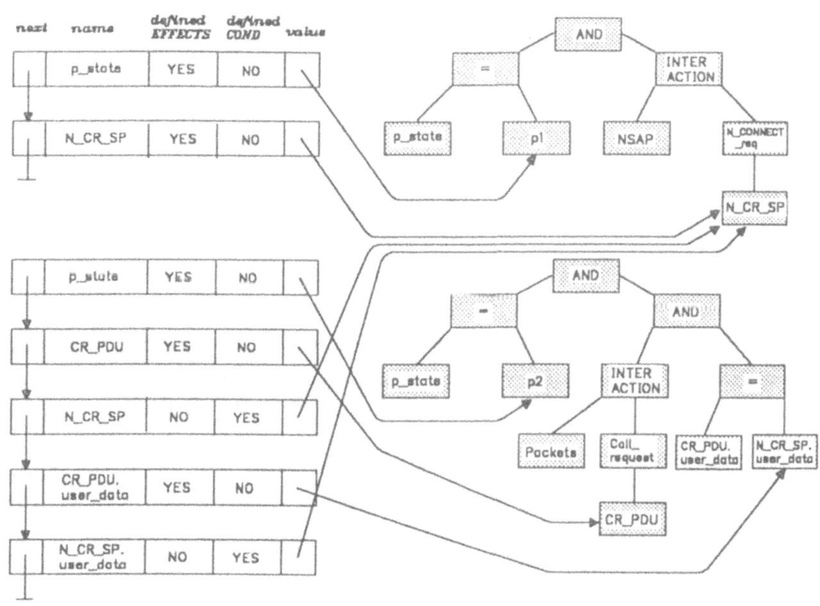

Figure 91. Dependency graph example for conjunctions

rule with several ∨-connected conjunctions), the dependency graphs for all conjunctions are connected to form the graph for the whole formula in DNF (see Figure 92). After analyzing the variable dependencies, the code for the five different functions implementing a rule is generated based on the dependency graph. If variables remain without a defined occurrence in neither $F_{cond\,(rs,\,r)}$ nor $F_{effects\,(rs,\,r)}$, no code generation is possible for this rule. Defined occurrences that are found result in assignment statements, whereas the used occurrences result in expressions corresponding to the predicate in the literal.

As an example, the code generated for the functions of the rule n_con_req are given in Figure 93. All functions have two parameters holding pointers to the rule control block (r) and the rule system control block (pm) of the rule. Within the functions, the addresses passed via parameters are assigned to local pointers, referring to individually structured objects in memory. In the n_con_req_CST-function, the requirements on the actual

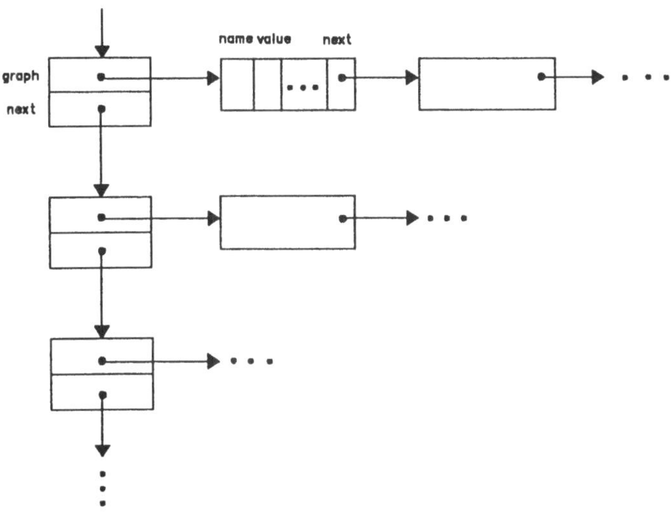

Figure 92. Dependency graph for disjunction

state of the rule system are evaluated, and in the n_con_req_CEV-function, constraints on event parameters are checked. If no constraints on event parameters are expressed in the start condition, or if a rule does not request an event, the function simply returns value 1 (true). The values found for the local variables are stored into the rule control block by the n_con_req_SLV-function (r->lvars). The values of the received event parameters are copied from the event control block (r->check_parms). In the n_con_req_EEV-function, the values of the offered event parameters are copied to the event control block (r->parms), and in the n_con_req_EST-function, the new values of the state objects are assigned to the rule system control block (pm->pstate).

The code generated for rules in such a *compilation* approach is deterministic and has been statically derived from the rule specifications. All variable dependencies have to be resolved after the analysis and the defined occurrences that are found uniquely determine the values of the variables. Specifying nondeterminism by allowing a range of values is not possible if the specification is to be compiled into executable form automatically. In this case, the analysis of dependencies fails in the tool

```
int n_con_req_CST (PRrepr r, PRsrepr pm)
{
   PX_25_DTE_state pm_state;

   pm_state = (PX_25_DTE_state) pm->pstate;
   return(*pm_state->asn_p_state == p1);
}


int n_con_req_CEV (PRrepr r, PRsrepr pm)
{
   return(1);
}


int n_con_req_SLV (PRrepr r, PRsrepr pm)
{
   N_CONNECT_req_ev    * pev;
   PX_25_DTE_decl          pl;

   pev = (N_CONNECT_req_ev *) r->check_parms;
   pl  = (PX_25_DTE_decl) r->lvars;
   *pl->asn_N_CR_SP = *pev->asn_n_connect_req;
   ...
   return(1);
}


int n_con_req_EEV (PRrepr r, PRsrepr pm)
{
   Call_request_ev     * pev;
   PX_25_DTE_decl          pl;
   PX_25_DTE_state        pst;

   pev = (Call_request_ev *) r->parms;
   pl  = (PX_25_DTE_decl) r->lvars;
   pst = (PX_25_DTE_state) pm->pstate;
   ...
   *pev->asn_call_request.user_data = *pl->asn_N_CR_SP.user_data;
   ...
   return(1);
}


int n_con_req_EST (PRrepr r, PRsrepr pm)
{
   PX_25_DTE_state pm_state;

   pm_state = (PX_25_DTE_state) pm->pstate;
   *pm_state->asn_p_state = p2;
   return(1);
}
```

Figure 93. Example rule functions

described in [Scho90]. However, nondeterministic behavior can still be expressed and made executable using different rules, or different rule system execution sequences. Taking an *interpretation* approach allows the dynamic evaluation of formulas at runtime and simplifies the handling of structured variables, but is not so efficient. However, depending on the intended use of the executable specification, interpretation might be more suitable.

Compilation of ADTs

The compilation of ADT specifications works similar to rule systems. In general, the task is simpler due to the following reasons:

- ADT objects, unlike rule system instances, are passive objects and their operations are invoked from outside the ADT;
- ADT operations are specified by a precondition (PRE) and effects (EFFECTS), but, unlike rules, their execution is atomic;
- ADT objects are defined using a number of state objects, but, unlike rule systems, there are no local objects used by rules.

The code generation for ADT specifications also starts from the attributed syntax tree, where all formulas used in operation specifications have been expanded and normalized to DNF. In the header file generated from the ADT specification, a type is defined for the state objects of the ADT. An example is given in Figure 94 for the ADT `DataQueue` specified in chapter 3 (see Figure 62). The X-ASN.1 type definitions required for the state objects are included by a corresponding preprocessor directive (part I in Figure 94). Then, a type definition summarizing the state objects is generated (II). The ADT 'copy'-function and procedures for all ADT operations are exported to other modules by external function declarations contained in the ADT header file (III, IV). Finally, an external declaration for the 'create'-function of the ADT is given (V). The implementation of these functions is contained in the source file for the ADT.

In the source file generated from an ADT specification, the header file is included (see part I of the example given in Figure 95). Then, the implementation of the ADT 'copy'-function follows (II), which works in a similar way to gates and rule systems. Every operation defined in the ADT specification is translated into a single procedure checking the precondition and

```
//-- I --------------------------------------- X-ASN.1 preprocessor directives --
$add_asn_type_module  ASN1_STANDARD_TYPES
//-------------------------------------------------------------------------- I --

//-- II ------------------------------------- state variable type definitions --
typedef struct {
   size INTEGER;
   ...
   } DataQueue_state;

typedef DataQueue_state *PDataQueue_state;
//------------------------------------------------------------------------- II --

//-- III ---------------------------- external declaration for COPY-function --
extern int DataQueue_STATE_COPY (Gen_pointer from, Gen_pointer *to);
//------------------------------------------------------------------------ III --

//-- IV --------------------------- external declarations for ADT operations --
extern int DataQueue_init (PDataQueue_state pstate, INTEGER new_size);
 ...
//------------------------------------------------------------------------- IV --

//-- V ---------------------------- external declaration for CREATE-function --
extern PDataQueue_state CR_DataQueue ();
//-------------------------------------------------------------------------- V --
```

Figure 94. Example ADT header file

performing actions to establish the specified effects if the precondition is satisfied. As an example, the `DataQueue_init` procedure is shown in Figure 95 (`III`). The parameter list of the procedure is the same as in the specification of the operation. The translation of ADT operations into a single procedure is possible because of the atomicity of operations. Local variables and event parameters do not exist for operations, declaration variables denote operation input and output parameters. A 'create'-function is generated for each ADT in the source file (see part `IV` in Figure 95), serving for the creation of ADT instances from within other modules. The 'create'-function uses the 'copy'-function to allocate memory and to assign the initial values to the state objects. During dynamic execution of the specification, the procedures implementing the ADT operations are the only functions that are invoked on ADT objects.

```
//-- I ------------------------------------------ include ADT header file --
#include "dataqueu.h"
//------------------------------------------------------------------- I --

//-- II ---------------------------------------------------- COPY-function --
int DataQueue_STATE_COPY (Gen_pointer from, Gen_pointer *to)
{ ... }
//------------------------------------------------------------------ II --

//-- III ----------------------------------------- ADT operation procedures --
int DataQueue_init (PDataQueue_state pstate, INTEGER new_size)
{
   if (TRUE) {
      *pstate->size  = new_size;
      *pstate->front = 1;
      *pstate->rear  = 1;
      *pstate->empty = TRUE;
      *pstate->full  = FALSE;
   }
   return(1);
}
 ...
//----------------------------------------------------------------- III --

//-- IV ---------------------------------------------------- CREATE-function --
PDataQueue_state CR_DataQueue ()
{ ...
  rc  = DataQueue_STATE_COPY (NULL, &state_area);
  ...
  return(state_area);
}
//------------------------------------------------------------------ IV --
```

Figure 95. Example ADT source file

Compiling the configuration

Besides modules for each rule system, gate, and ADT specification, another module is generated for creating the rule system and gate instances used in the CRS specification and for initializing the data structures of the scheduler. The 'create'-functions exported by the individual modules are used to accomplish this task.

In Figure 96, the code generated for the configuration specified in Figure 46 (see chapter 3) is shown. The 'create'-functions which are required are imported from other modules by including the individual header files (I). Then, the gate instances are created and inserted into the *scheduling queue for gates* (GATEQUEUE) managed by the scheduler (II). With the gate instances available, the rule system instances are created

```
//-- I ---------------------------- include gate and rule system header files --
#include "n_sap.h"
#include "x_25_acc.h"
#include "x_25_dte.h"
//------------------------------------------------------------------------ I --
void initialize_system (void)

{ PSynGateset nsap1, packets1, ... ;
  PRsset      ..., network_entity1, ... ;

//-- II -------------------------------------------------- create gate instances --
nsap1    = CR_Network_service_access_point ("NSAP1");
INS_SGATE (&GATEQUEUE, nsap1);

packets1 = CR_X_25_network_access_point ("Packets1");
INS_SGATE (&GATEQUEUE, packets1);
 ...
//----------------------------------------------------------------------- II --

//-- III ---------------------------------------- create rule system instances --
 ...
network_entity1 = CR_X_25_DTE ("Network_entity1", nsap1, packets1, 1);
INS_RS (&SCHEDQUEUE, network_entity1);
 ...
//---------------------------------------------------------------------- III --

return(0);

}
```

Figure 96. Example configuration source file

and inserted into the scheduling queue for rule systems (SCHEDQUEUE) managed by the scheduler (III). All parameter values that are required to create the instances are given in the specification. The procedure initialize_system is called by the runtime environment before the dynamic execution of the specification is started.

5.2 CRS Integrated Tools Environment

In order to apply the development methodology outlined in chapter 1 powerful tool support has to be provided. A variety of tools supporting different development steps were identified in section 1.3. The realization of these tools is facilitated by the formal syntax and semantics of FDTs employed in the development process, as described in chapter 4. Many of these tools are based on specification interpreters.

Apart from the design and realization of each single tool, the integration of these tools for all development activities into a common environment is necessary in order to support the whole development process. As illustrated in chapter 4, many existing protocol development tools focus on a single activity and are only loosely coupled with tools for other activities in existing environments. Very often the capabilities of an FDT simulator are used to accomplish additional tasks besides design validation, for example, trace analysis, test generation etc. However, since these additional tasks were not included in the design of the simulators, flexibility is sometimes severely limited in such tools.

In the following, the general architecture of an integrated tools environment is presented that combines protocol engineering tools in an efficient and flexible way. The idea is to support the formal development process described in chapter 1 in an environment comprising tools for each development activity. The interworking of these tools is facilitated by letting them share common information derived from the formal descriptions. In order to support this idea by a suitable architecture a logical separation of the environment into three different layers is made: kernel, shells, and application tools. A realization of these layers on the basis of the X-ASN.1 and CRS implementation concepts and tools explained in section 5.1 is described.

5.2.1 Architecture

The tools identified in chapter 1 (see section 1.3) can be divided into two classes:

- **system tools** supporting a formal technique itself, focussing on the specific formalism and not on its application. Examples are specification language editors, compilers, interpreters, simulators, etc.
- **application tools** supporting the application of a formal method to achieve a particular purpose. Examples are validation tools, verification tools, test suite generation tools etc.

Application tools support development activities, which are independent of the particular FDTs used, whereas system tools are tailored towards specific techniques and provide the means for realizing this support. System tools should offer a comprehensive set of flexible mechanisms, even permitting contrary strategies to be employed in application tools. For example, system tools for interpretation and simulation may be used by different application tools. This separation is crucial for the integration of several tools into a common environment: system tools are provided in a kernel and offer the entire set of their capabilities to a variety of application tools (**functional integration**).

Another important integration aspect is the following: tools used in all the phases of communication system development require and produce knowledge about the services or protocols they are working on. This knowledge may be represented differently, together with corresponding storage, retrieval, and inference facilities. State tables, attributed syntax trees, reachability graphs, behavior expressions, predicates, and rules are just a few examples. Therefore, the second basic idea of the integrated environment architecture is that protocol knowledge is shared by different tools (**knowledge integration**).

The architectural concept proposed for an integrated tools environment takes both of the above considerations into account (see Figure 97). The environment **kernel** comprises all the system tools for the formal techniques to be used and contains the knowledge base for the communication system under development. The **shells** around the kernel offer flexible access to the system tools and provide flexible views on the knowledge base. This may be a state-oriented view or an event-oriented view, focussing on reachability, external observability, parameter dependencies, or desired properties. The knowledge represented in the kernel and the inference mechanisms provided are utilized to offer different conceptual interfaces to the outermost layer of the environment, formed by various **application tools**.

The shells abstract from the particular techniques and representations employed in the kernel. This takes into account that each application tool has a preferred system specification and solves the conflict between state-orientation and event-orientation. If, for example, the formal technique employed in the kernel is state-oriented, application tools working on event sequences may still be integrated into the environment through an event-based shell. Application tools may combine several shells to accomplish the tasks of formalization, validation, stepwise refinement, implementation,

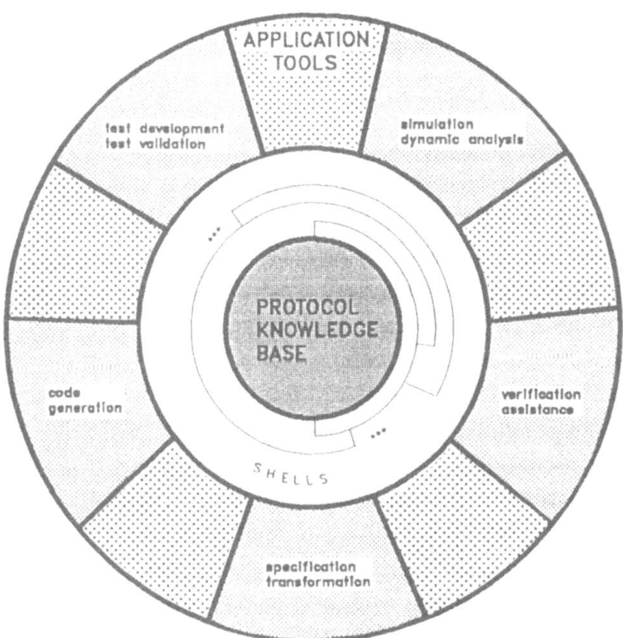

Figure 97. Integrated tools environment

and testing, depending on the degree of abstraction required and the strategies and heuristics used. Such an integrated environment has considerable advantages:

- protocol information is consistent for all tools;
- modifying the protocol under development is done centrally;
- individual application tools are easier to develop;
- interworking of several tools is possible;
- the environment can easily be extended.

In the following sections, a realization of the three basic layers of such an integrated environment is outlined, based on X-ASN.1 and CRS employed in the kernel. A *rule-based*, *state-based*, and event-based shell are described.

How these shells can be utilized differently is illustrated by investigating some selected application tools.

5.2.2 Kernel

In the kernel, information about the communication system under development has to be represented and processed. The information about services and protocols is given by formal descriptions in textual or graphical form. This *external format* has to be translated into an *internal format* by a set of system tools. The goal is to represent the protocol knowledge in an adequate way and to provide mechanisms for knowledge management and retrieval.

The formal descriptions contain different kind of information. The structure, contents, and encoding of protocol data is one aspect. Protocol data structures, data instances, and operations on data items can be described by means of X-ASN.1 and are managed internally by a corresponding data management component in the kernel. Protocol behavior, in terms of external visibility, is another aspect. Behavioral aspects can be adequately described by using CRS rule systems that are represented internally as production systems. An X-ASN.1 and CRS implementation has been described in section 5.1. A visualization of the kernel based on these implementation concepts Is given in Figure 98.

It has to be noted that additional information about a service or protocol concerning non-functional aspects, for example, performance, reliability, security, resources, exceptions etc. cannot be described in CRS or X-ASN.1. Appropriate constructs and a corresponding semantics for specifying these aspects still have to be developed. Consequently, there are presently no means for getting the non-functional information into the X-ASN.1 and CRS based kernel. Extensions for capturing non-functional aspects in FDTs, especially performance aspects, are currently the subject of intensive research, but are outside the scope of this thesis. How data and behavioral aspects are represented and processed in the X-ASN.1 and CRS based kernel is further detailed in the following.

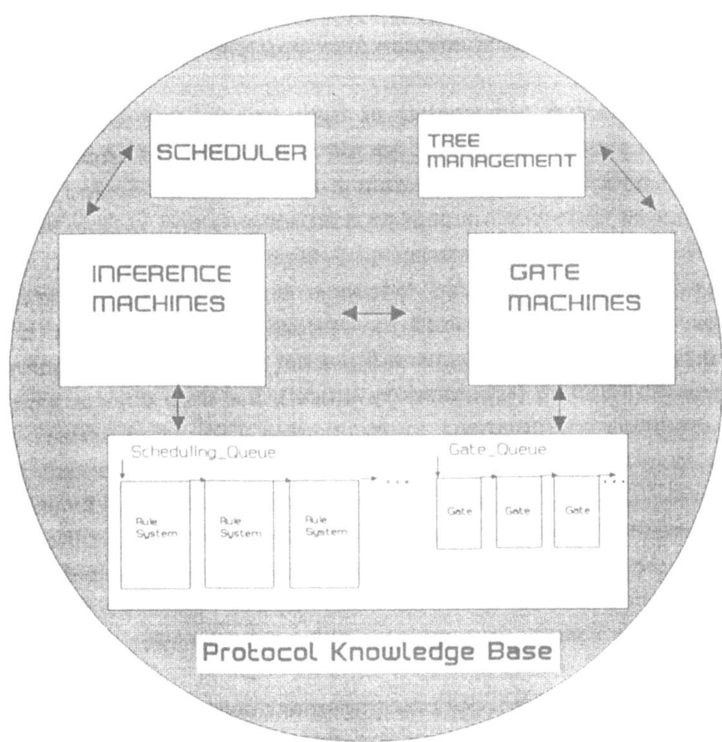

Figure 98. CRS and X-ASN.1 based kernel

Data aspects

A large number of concepts have been developed for representing data in systems. Despite many differences, all concepts introduce a clear separation between data structures (or data schemes) and data items (instances of a certain structure). A data management component has to provide structuring mechanisms, as well as storage and retrieval mechanisms for data items [Kort86].

Data schemes for SPs and PDU formats can be created by using the X-ASN.1 type definition constructs explained in chapter 3. The **X-ASN.1 compiler** described in section 5.1 checks syntax and static semantics, resolves external references, and converts the type definitions into the internal **structure tree** representation used by the X-ASN.1 data management component.

For each type module the global symbol and index tables are built, serving for the retrieval of type information from structure trees.

Data instances are represented as **data trees** composed of data blocks. System tool support is provided for either interactive or embedded creation of data instances. An **X-ASN.1 data tree editor** can be used to allocate a new instance of an X-ASN.1 defined data scheme and to fill in the values for the different components interactively. Such a tool is described in [Sche89], providing a window-oriented interface and performing validity checks during input. It can be utilized for interactive protocol testing, or for executing specifications of systems that are not closed, but share gates with the external environment (simulated by a user). The state objects, local objects, and parameters embedded in executable CRS specifications are data instances that are created by runtime routines (for example, during protocol validation). The corresponding function calls are inserted by the **X-ASN.1 preprocessor**. The **tree management functions** provided by the data management component allow the access, retrieval, and manipulation of data trees. Encoding/decoding of data instances is achieved by calling the generic **X-ASN.1 encoder/decoder** working on data trees.

In summary, the X-ASN.1 data management component in the kernel provides a set of functions that can be used in various ways by application tools supporting different development activities. In order to cope with the large amount of data being managed, for example, during dynamic state exploration or derivation of a large number of test cases, data trees may be stored on disk in an external format and loaded again at some later stage.

Behavioral aspects

A communication service is defined by a set of SPs exchanged between a service provider and a service user, possible sequences of SPs, and parameter dependencies. This view focuses on the externally visible behavior of a communication system and abstracts from the internal behavior. How a service is provided by a number of distributed entities is described in a protocol specification. The protocol determines the behavior of each entity: the set of possible input events in each state, the actions to be executed in response to various inputs, and the corresponding output events.

Using CRS, the relation between input events and output events, possible event sequences, and possible event parameter ranges are described by rules defining the behavior of rule systems. An internal state concept is used to facilitate the description of sequencing aspects. Several rule systems can be composed in parallel and cooperate to provide a service. The behavioral information contained in CRS specifications can be represented in form of **production rules** in the kernel, as explained in section 5.1. Besides formal logic, procedures, and frames, production rules are a basic technique for representing knowledge in systems [Barr81] [Fros86]. In the production systems obtained with the **CRS compiler** described in section 5.1, a mixed form of representing knowledge about behavioral aspects is used. Whereas production rules capture the input/output relations and spontaneous actions, **rule body procedures** represent the knowledge about internal state conditions and event parameter constraints.

A procedural form of representing knowledge is appropriate for simulation and implementation, but sometimes a declarative representation is more suitable, for example, for verification or test case development. Therefore, formal logic employed for specifying rules in CRS may also be used for representing protocol knowledge. Similar to data types and data instances, the start condition and effects formulas of rules can also be represented in the form of trees (so-called **constraint trees**). These constraint trees are similar to structure trees and data trees, but contain symbolic information about predicates instead of data values in leaf nodes. This is an alternative representation of state and parameter constraints better suited for the retrieval of information on particular state objects and parameters as opposed to compiled procedures, which deliver the truth value of start condition and effects formulas without any individual information on state objects and parameters.

The **runtime environment** in the kernel provides the generic machines to work on the protocol knowledge and to offer knowledge retrieval, inference, communication, and scheduling functions. These kernel mechanisms can be used by application tools to analyze behavior sequences, generate reachability graphs, check invariant conditions, reason about desired properties, or perform various other protocol engineering tasks.

5.2.3 Shells

Experiments with the first prototype implementations of the X-ASN.1 and CRS system tools (see [Schn87] [Schn89a]) clearly showed that providing flexible access to the environment kernel via appropriate interfaces is of utmost importance for building application tools. The implementation of an *application program interface* provided in a function library interfacing with kernel routines was a major step towards a comprehensive integrated tools environment. This application program interface constitutes the **rule-based shell** and is described in more detail in the following.

However, the rule-based shell is not adequate for application tools working on higher levels of abstraction, for example, for tools working on externally observable behavior sequences or global states. Therefore, two additional shells providing complete transparency of the techniques used in the kernel are described in this section: the **state-based shell** and the **event-based shell**. By combining functions from different shells, application tools can use different views of the specified system in an interleaved fashion. The shells act as bridges between different models and allow advantage to be taken of one model while avoiding the disadvantages of others.

Rule-based shell

Figure 99 illustrates the application program interface constituting the rule-based shell of the X-ASN.1 and CRS based kernel. The functions provided are tailored towards the techniques and work directly on the internal representations of data objects, rule systems, gates, and rules [Schn89b]. The set of functions provided by the rule-based shell can be divided into the following groups:

- **Logging** functions, which control the data logged during system execution. These functions may be called at any time to increase the amount and granularity of information that is written to disk for a subsequent analysis, or to decrease the level of detail to speed up execution when no interesting events are expected. The scheduling of rule systems and gates, the selection of rules to be started or ended, the events to request and offer, the transitions of rules between sets, state changes, and other kinds of information may be logged while working with the system.

- **Rule system** functions, which return information about selected state objects, contexts and views defined in the specification, running and non-running rules, and rule start conditions and effects formulas. Functions are provided in this group which can be used by application tools to determine the set of startable and endable rules in a given state of the rule system.
- **Gate** functions, which return information about the state of gates and the available event requests and offers. There are also functions provided in this group which can be used to add or remove event requests and offers at external gates.

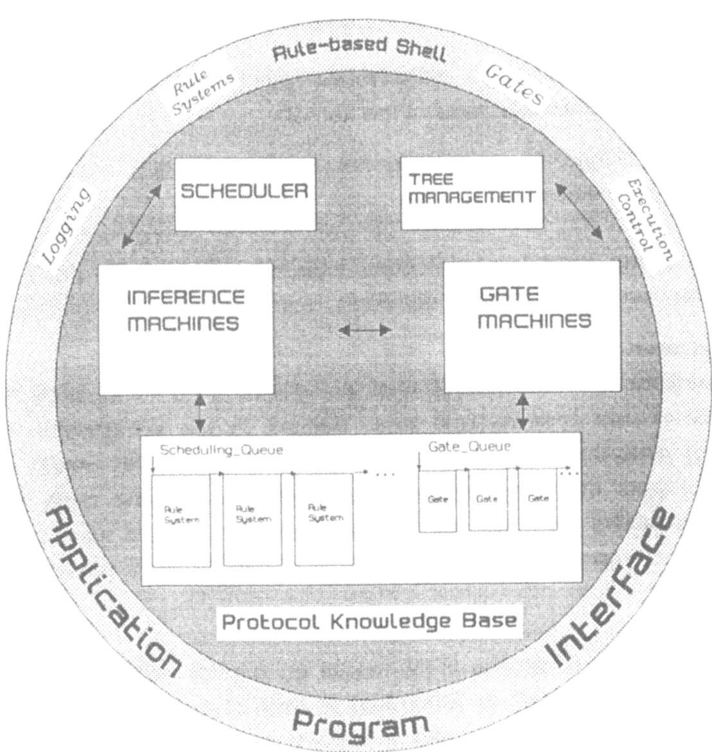

Figure 99. Rule-based shell

- **Execution control** functions, which can be used to invoke system actions, for example, scheduling rule systems and gates, or starting and ending rules. Whereas functions from the other groups give information about possible alternatives, but do not change the current state of the system, the functions from this group bring the system into one of its possible next states.

For most of the functions of the rule-based shell, some mechanism for identifying specification objects has to be used. The names of rule systems, rules, and gates suffice to interact with application tools, but more efficient approaches can be taken as well. Also, structure trees, data trees, and constraint trees can be exchanged via the application program interface. The rule-based shell is appropriate for application tools working directly on rules and other objects that are specific to CRS and X-ASN.1. Therefore, important data structures managed in the rule-based shell are:

1. *structure trees* and *data trees*, *control blocks* and *scheduling queues* (internal structures of the kernel);
2. *logging flags*, for controlling the logging activities individually for a number of interesting aspects.

Besides being used by application tools, the functions of the rule-based shell may be used by additional shells, as explained in the following.

State-based shell

The functions of the rule-based shell are mainly used for retrieving information about rule systems and rules defined in the specification and for invoking state transitions by starting and ending rules. No means are provided to work on global states of the system, or to step back and take some alternative execution path after having executed a number of rules. This kind of support is provided by functions of the state-based shell (see Figure 100).

Since CRS is based on the EFSM model an explicit notion (and representation) of states is given by the values of state objects, the execution states of rules and local objects, and the event offers available at gates. The internal representation consists of data trees and control blocks. For man-

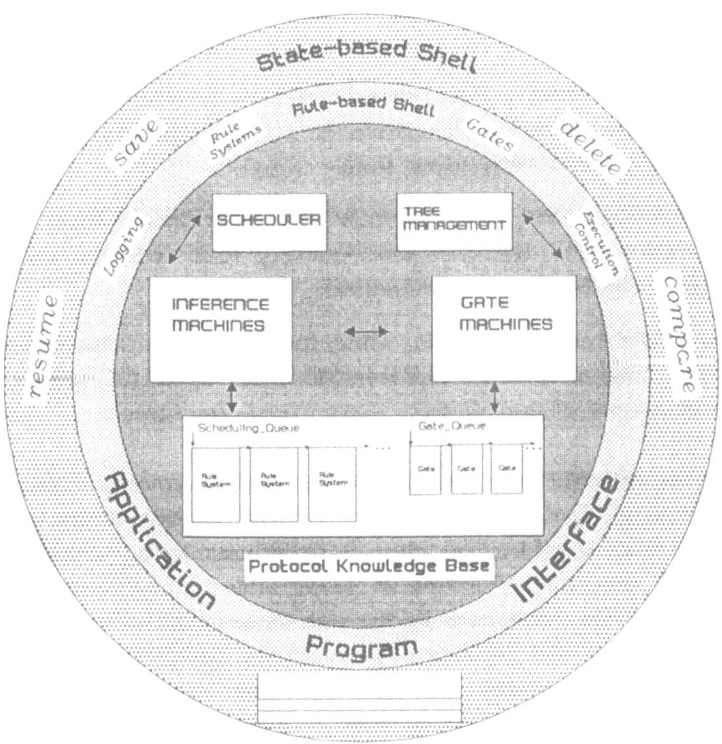

Figure 100. State-based shell

aging a large number of states in the state-based shell, data structures have to be maintained for

1. storing states (referred to as *system stack* in the following);

2. associating states with identifications (*state identification table*).

The following functions are provided by the state-based shell:

- The function **save_state**, which creates a copy of the current state of the system on the system stack. Copies of the control blocks for rule systems, rules, gates, event requests, and event offers are required, as well as copies of data trees representing state objects and local objects.

An identification is returned by the function, providing a reference to the saved state.

- The function **resume_state**, which restores a previously saved state and resumes the execution from there. All control structures are reset to the values of the previously saved state.
- The function **delete_state**, which removes a previously saved state from the system stack. The associated memory space can be released and the state identification can be reused.
- The function **compare_state**, which takes two state identifications and checks if the associated representations are equal (all variable values are equal and all control block information is the same)[11].

Other functions required in the state-based shell are mainly state management functions, serving for searching, moving, copying single states etc. Complete search and backtracking strategies have to be realized by application tools according to individual needs. State management and organization is, however, taken over by the state-based shell. In turn for the extra effort of implementing backtracking algorithms, application tools can flexibly adjust backtracking strategies to their needs. Experience with "PROLOG", for example, shows that a predefined fixed backtracking strategy becomes inappropriate with growing search spaces.

Event-based shell

A lot of application tools work only on the basis of the externally visible events of a system and do not use any details of the internal model of a service or protocol specification. Such application tools should access the kernel mechanisms via an event-based shell. For the X-ASN.1 and CRS based kernel such an event-based shell has been implemented on top of the rule-based and state-based shells as described in [Roor90] (see Figure 101). The functions provided by the event-based shell are the following:

[11] Note that this is an equivalence on states which is rather strong with respect to externally visible behavior.

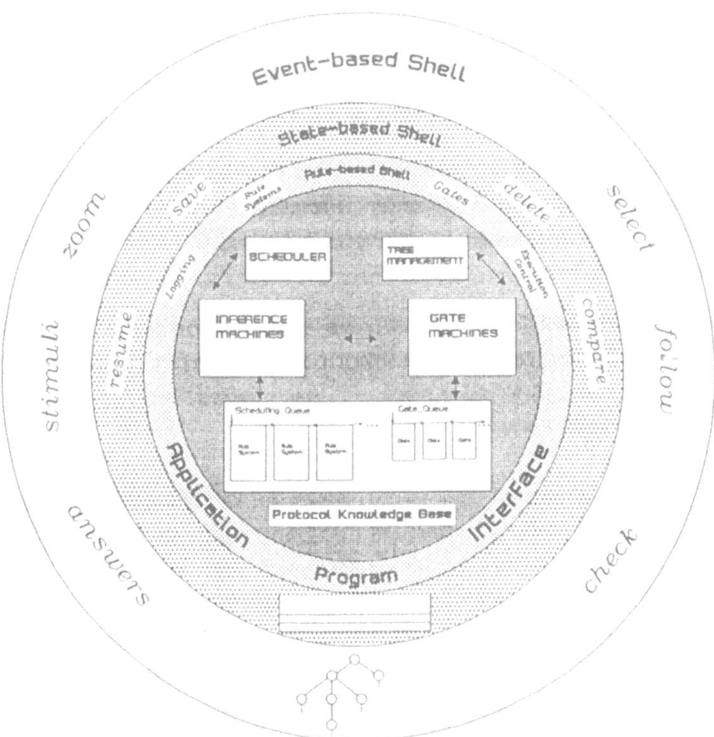

Figure 101. Event-based shell

- The function **zoom**, which restricts further attention of the event-based shell onto a set of events that are considered interesting by the application. In this way, the scope of subsequent searches is limited. The focus on a particular set of events can be modified by subsequent calls of the function.

- The function **stimuli**, which lists all events that can be accepted by the system in a given state at the external gates. Gates serving for internal communication between system parts are not considered.

- The function **select**, which marks one of the possible stimuli as being selected for subsequent behavior. This function is called for making a choice, but it does not yet invoke any actions.

- The function **answers**, which returns all possible responses after some specific selection. The list of events returned as the result of the function contains all events that can be offered by the specified system in the state reached after the selected event has been accepted.
- The function **follow**, which instructs the executable specification to progress the system execution according to one of the possibilities returned by the answers-function. In this way, the system reaches a new state.
- The function **check**, which can be used to compare some presumed behavior with the possible behavior according to the specification. Given two events, the function checks if the second event is an admissible answer to the first event.

In order to find out which events can be accepted (offered) by the specified system, rules have to be started (ended) and states have to be saved and resumed (using functions of the rule-based and state-based shells). The stimuli-function, for example, searches for startable rules that request events at external gates in their start condition. The answers- function has to start each rule requesting the selected event separately from the current state. Then, one of the three following cases applies: (1) the started rule offers an event at an external gate, (2) the started rule offers an event at an internal gate, (3) the started rule offers no event in its effects. In case (1), a possible answer has been found. In case (2), matching event requests from startable rules in other rule systems have to be found, these rules have to be started, and the procedure is applied recursively to all states reached. In case (3), all startable rules without event request in the same rule system have to be started and the procedure has to be repeated for every state reached. However, all this internal processing is hidden from the application tools using the event-based shell.

In order to implement the functions described above, the event-based shell maintains two data structures, called *event info* and *tree info*, which hold the following information:

1. *Event info* contains the list of external gates in order to be able to distinguish between requests/offers made at internal and external gates. Sim-

ilarly, lists of actually zoomed and not-zoomed events are maintained. Finally, the event identified by the selection-function has to be marked in the list of zoomed events.

2. *Tree info* is a representation of the behavior tree associated with the specified system. Each edge in the tree is labelled by an event. Each node corresponds to a state and contains an identification and additional information about the rule causing the transition to that state.

In addition to the possible event sequences, the data carried in event parameters is important. For example, when establishing a connection, properties of the connection may be negotiated through fields in the CONNECT and ACCEPT PDUs. These fields are modelled as event parameters. The constraints defined on them are important information for application tools. Therefore, the event-based shell has to provide functions that can be used to query the constraints on event parameters and to check a particular event with all its parameters for validity. To realize these functions, the constraint tree concept explained in the previous section can be used. The stimuli-function, for example, can be extended to return constraint trees for all the events that can be accepted in the current state. Concerning responses, constraint trees represent possible parameter ranges and are returned by the answers-function. To progress the executable specification into another state, an answer is chosen and the follow-function is called with appropriate parameter values.

5.2.4 Selected Application Tools

In this section, the design and realization of some selected application tools for protocol development is illustrated, starting from the different shells presented in the previous section. The selected tools support the activities of protocol validation and testing by interactive behavior simulation, automated test case development, and test case validation against a formal description. The experience gained with the implementation of these application tools while preparing this thesis is the main reason for including an outline of their design in this section. The goal is to illustrate how interactive design validation and testing can be effectively supported by automated tools. In general, the architecture of the integrated tools environment is flexible enough to build tool support for other development activities as well (verification of properties, stepwise refinement by automated transformation, implementation code generation etc.).

Interactive behavior simulation

The CRS simulator gives a protocol engineer the opportunity to analyze an executable version of the communication system design in an early phase, at varying levels of speed and detail. In the beginning, attention can be restricted to the events occurring at the service interfaces. Service users can be modelled as rule systems, or can be simulated by the user requesting and offering events at gates representing SAPs. Later, the behavior of protocol entities can be checked by looking closer at PDUs and the communication at the internal gates.

The availability of rule-based and state-based shells around the X-ASN.1 and CRS based kernel (containing the internal representation and the runtime environment) simplified the development of a specification simulation tool considerably. The remaining task to be accomplished was to provide a flexible and user-friendly user interface for displaying information appropriately and for offering convenient means to invoke desired actions.

The top-level windows of the CRS simulator described in [Schn88] are depicted in Figure 102 for an executable specification of the rule system configuration specified in chapter 3 (see Figure 46). The user of the tool is able to create this initial configuration of rule system and gate instances, to associate a window with each instance, and to position the windows on the screen. A subset of the state objects can be selected for each instance and the values of the selected objects are displayed in the corresponding windows.

Function keys allow zooming into selected rule system and gate instances in order to get more detailed information about rules, or event requests and offers. The logging flags can be set appropriately to trace the simulated behavior. Using the menu bar, scheduling and inference machine strategies can be adapted. The execution mode can be set to follow predefined strategies individually for each rule system and gate instance at the convenience of the user. For each X-ASN.1 defined object the X-ASN.1 data tree

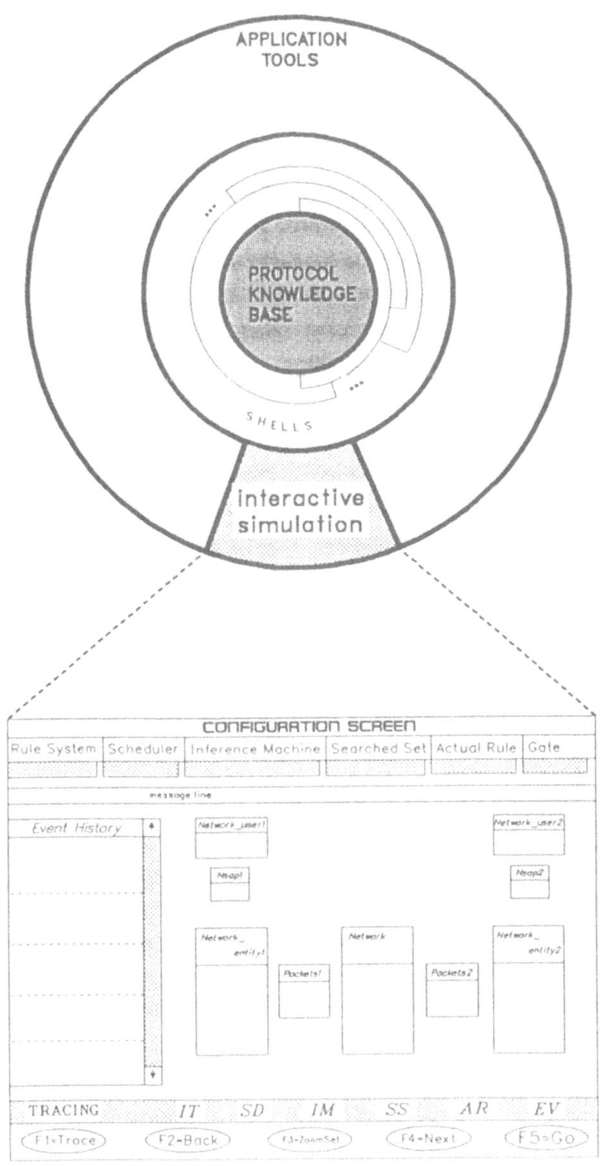

Figure 102. Behavior simulator

editor can be called in order to examine data trees in detail, to modify values if desired, or to create user-defined event requests and offers.

After scheduling a rule system, the startable or endable rules in the current state of the rule system are listed in a menu. By making selections from this menu, rules can be started and ended, which brings the whole system into a new state. The backtracking features of the state-based shell can be used to resume the simulation from previously saved states. The events occurring during the simulation are displayed and stored in an event stack. The user may also add event requests and offers at external gates for triggering actions in the executable specification.

With the control facilities described above, a protocol engineer can easily influence the execution of the specification and drive it to areas of interest. More or less detailed traces may be produced and analyzed by other tools after the simulation. The experiences made with a prototype of the CRS simulator showed that such a tool is a powerful instrument to validate a design for the most interesting situations and that it is also useful for teaching services and protocols to newcomers. However, the execution mechanisms provided may also be used to set up long-term automatic simulations. Several strategies may be employed (full reachability, modified reachability, random state exploration etc.).

Test case development and execution

Testing protocol implementations is necessary to achieve reliability and to guarantee interoperability of open systems. The goal is to increase confidence in the correctness of an implementation and, thus, to increase the probability that the implementation is able to interwork with other implementations of the same protocol. In the context of open systems, testing is aimed at checking the externally visible behavior of implementations and their conformance to standards (*conformance testing*). The *implementation under test (IUT)* is treated as a black box exchanging primitives at its external interfaces according to the test configuration used (see for example Figure 65 in section 4.2.3).

Because of the large number of implementation options and possible combinations of primitive parameters existing for real protocols, testing can never be exhaustive. The production of test cases and their aggregation into a complete test suite covering as many protocol aspects as possible, as well as the related questions about test purpose, test selection, test quality, test coverage, and test execution are major research topics in protocol engineering today. Considering the variety and complexity of today's and future protocols, tool support for test case development becomes increasingly important since such tools speed up the production of tests and help to improve test suite quality.

In order to automate test case development, a formal description of the protocol to be tested is required, providing all the information about correct and incorrect protocol behavior in a form which can be processed by tools. However, fully automatic production (generation) of test suites covering all protocol aspects is difficult and cannot be achieved by any of the methods summarized in section 4.2.3. The test generation tools currently known using FSM descriptions of the protocol generate tests for each single transition in the underlying state machine, but do not consider complete behavior sequences, nor do they take any data aspects into account. However, data aspects become increasingly important and complex in the higher layers of the OSI model, whereas the control aspects become less important. Test generation methods integrating data aspects use static data flow analysis and other techniques developed for software testing. However, all formal strategies suffer from the fact that it is not possible to include information about the functions expected from a particular protocol in the test derivation algorithm. Consequently, the generated tests may be more or less appropriate, depending on the particular protocol to be tested. A test derivation algorithm delivering *canonical testers*, which represent test suites for every possible behavior defined in the specification, shifts the complexity of test generation to the activity of test selection.

For all these reasons, a test case development tool (instead of an automatic test case generation tool) operating in an interactive fashion with a test case designer in the integrated tools environment is considered to serve existing needs. A first prototype of such a tool has been implemented according to the configuration shown in Figure 103 (see also [Velt89]). Test case design is conducted on the basis of an executable reference specification of the protocol, excluding the possibility of errors and providing all

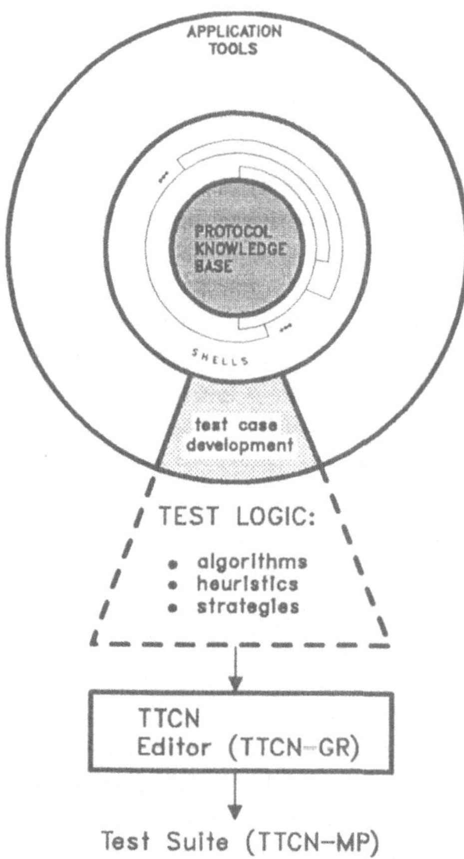

Figure 103. Test case development tool

necessary information about the protocol to be tested. The user of the tool retrieves protocol information from the kernel using functions of the event-based shell.

With the zoom-function, he can focus on a number of test events that are interesting for achieving a specific test purpose. Then, the stimuli-function can be used to list the test initiatives together with their parameter constraints. One event can be selected with the select-function and a combination of parameters can be supplied. By calling the answers-function, the possible responses of the protocol as defined in the specification can be determined and the test case designer can invoke the follow-function with one of the alternatives.

The functions of the state-based shell can be used to restore previously saved states in order to follow one of the other alternatives for developing another test case. At the end of each test path, verdicts can be assigned based on the protocol information retrieved from the environment kernel. The test case designer can be supported by different automated test derivation algorithms (based on FSMs, EFSMs, data flow etc.) implemented in the test development support tool. Proposals for the continuation of event sequences or for interesting parameter combinations can be made by the tool. In order to assess test suite coverage and test suite quality, statistics about states, transitions, tested events, parameter combinations etc. can be recorded.

The output of the test case development tool is displayed in standard TTCN format (see section 4.2.3) using the TTCN editor described in [Baer89]. A test case in TTCN consists of a tree of *test events* (see [ISO9646]). The events in each path of the tree starting from the root and ending in some leaf node occur in sequential order; events at sibling nodes represent test alternatives. Each test event is of the form

```
PCO !|? Event_id (variable_assignments) [boolean_expressions] .
```

A `PCO` denotes the *Point of Control and Observation* where the event occurs. The `Event_id` is usually associated with an SP or PDU exchange and the exclamation mark (question mark) means sending (receiving) the event. Values may be assigned to test variables when the event occurs. A condition in the form of a boolean expression can be associated with each event (only if this condition is satisfied may the event occur). All declarations of PCOs, events, test variables etc., and the test case behavior tree are contained in tables of fixed formats. All information required for creating these tables can be retrieved from the protocol specification represented in the environment kernel.

Besides offering an editing facility, the TTCN tool described in [Baer89] is also capable of interpreting TTCN test cases. Together with all the mechanisms provided in the kernel, a test tool can be realized that has many advantages over existing test tools (see [Mack88]). Most of the existing test tools for protocols have been especially conceived to be applied to one particular protocol in a stack of protocol layers. The heart of such tools is often formed by an approved reference implementation of the protocol to be tested. Reusing the tool for other protocol layers is not possible. In contrast to this, the test tool described in [Mack88] is separated into one (or several) language dependent **test processing systems (TPSs)** and a language independent **test support system (TSS)**. In the TPSs, execution mechanisms for test languages (for example, TTCN) are realized. The TSS provides common services to the TPSs for handling data, managing PCOs, and connecting to several communication systems. This test tool design meets three objectives facilitating the general usage of the tool:

1. **protocol independence**,
2. **test configuration independence**, and
3. **communication system independence**.

Protocol independence is achieved by employing general purpose test processing languages in the TPSs. Test processing is based on an executable specification of the protocol used as the reference. The flexibility of retrieving information via different shells allows the identification of possible reasons for erroneous behavior in an IUT and the selection of *conformance resolution tests* in order to locate an error. Test configuration independence is achieved by directing interactions through PCOs managed in the TSS. These PCOs may be connected to an interprocess communication facility or to external communication channels. Independence of the communication system is reached by providing several communication system interfaces in the TSS.

This example of several tools interworking in the integrated environment illustrates nicely how common functionality is provided in the kernel and does not have to be replicated in every single tool. The example also shows the importance of organizing protocol development tools in such a way that they provide combined support. Especially in the area of testing, the separate standardization of protocols and their associated conformance tests introduces unnecessary interpretation problems and new ambiguities due to potential differences between the protocol and the test specifications. Such a situation cannot arise if tests are directly derived and validated based on the specification.

Test case validation

Assessing test suite correctness and test suite quality is the key motivation for an application tool supporting test case validation. Such a tool is currently under development in the integrated tools environment. Correctness of a test suite means that there is no IUT implemented according to the standard which does not pass the test suite. Quality of a test suite depends on coverage, error detection capabilities, and the time required to perform the test suite. This list is, however, not exhaustive.

Driven by vendors of communication hard- and software, standardization bodies (ISO, CCITT) have already started producing test suites for standardization (for example, FTAM, X.25 layer 3 etc.) using the test description language TTCN. These tests have been developed manually and distributed for review by protocol experts and TTCN experts in order to gain confidence in the correctness of the tests. Considering the large amount of test cases and the complex protocol functions that should be tested, validation tools are needed.

The review process for test suites includes checking the TTCN syntax and static semantics, as well as validating the correctness of the test suite with respect to the protocol standard the test suite has been developed for. Checking the syntax and static semantics of TTCN can be automated by employing parser tools accepting the TTCN language (see for example [Baer89]). Checking automatically whether a test case assigns 'pass' verdicts only if allowed behavior has been observed, 'fail' verdicts only if an error has been found, and vice versa, is a more complicated and yet unsolved problem for manually developed tests.

An essential prerequisite of such an automated validation of test descriptions against the corresponding standard is to have the knowledge of protocol experts, currently reviewing the tests manually, accessible in software tools. In the integrated tools environment, all information about the protocol for which a test suite should be validated can be represented in the kernel, based on a specification of the protocol in X-ASN.1 and CRS. The TTCN test description is parsed, loaded into the TTCN editor, and checked against the executable protocol specification by interpreting the TTCN constructs (see Figure 104). The approach is rather pragmatic as long as a formal TTCN semantics is still missing, but may be justified by relating TTCN to the FDTs employed for protocol specification (see for example [Sari88]), or to more basic models (for example, *labelled transition systems*).

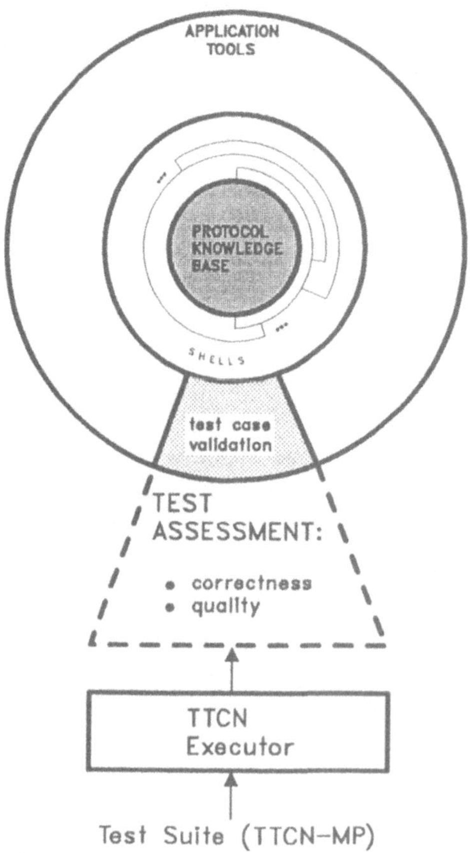

Figure 104. Test case validation tool

Since a TTCN test description consists of a tree of events, a possible strategy is to check every path in the tree using functions from the event-based shell. Admissible event sequences, valid parameter ranges, and behavior in the case of errors as defined in the specification are compared with the test description. Functions from the state-based shell are used to backtrack to intermediate states (or to the initial state) and to try the test alternatives. A major prerequisite for this approach is, however, that ASN.1 descriptions of test parameters and PDU formats are available. The use of ASN.1 is optional in TTCN, but simplifies automatic validation considerably. Future extensions of TTCN concerning "parallel" execution of test case behavior trees are necessary for performing *multi-party testing* (testing several entities executing, for example, an application layer protocol in parallel). For the FDTs employed in the kernel, constructs for describing parallel activities already exist.

While running the test suite validation tool, statistics can be maintained to gather information about which part of the protocol (service aspects, protocol aspects, state transitions, parameter combinations etc.) are tested thoroughly, and which parts are probably not tested at all. These statistics may be evaluated afterwards to allow qualitative and quantitative statements about test coverage (depending on whatever definition of coverage is used).

6 Conclusions

An approach to the complex tasks in protocol engineering aiming at higher quality of protocol implementations and reduced development costs has been presented. In order to achieve this goal the use of formal description techniques has been advocated and a suitable technique employing rules has been proposed (*Communicating Rule Systems* (*CRS*)). The application of CRS to an existing protocol of realistic complexity has been illustrated. Concepts and tools have been described facilitating the automatic processing of CRS specifications in order to support the protocol development process in an integrated environment. A summary of the main results, the status of the tool implementations, early experiences, and an outlook on future work is given in the following.

6.1 Summary of Results

Developing communication systems and implementing protocols on the basis of natural language specifications leads to a number of problems caused by ambiguities and misinterpretations. In such an informal approach, design validation, refinement, implementation, and testing have to be performed manually. Therefore, considerable effort is required and errors are likely to be made, especially since applications demand increasingly powerful communication services which have to be provided in heterogeneous environments.

The main benefits gained from applying FDTs in the protocol development process are preciseness and the ability to employ tools. The formal syntax and semantics of FDTs improves the understanding of the basic architectural concepts of communication systems and facilitates the provision of tool support. However, applying FDTs to the development of systems is not trivial. The formalization step is difficult and appropriate techniques have to be used in order to achieve a suitable formal description in reasonable time. Only if the technique correctly represents the basic architectural concepts of communication systems and offers constructs supporting intuitive and structured protocol specifications, it can be useful in practice. The feasibility of building tools that process the formal descriptions is crucial since tool support is the key motivation for the formal approach.

The investigation of several existing FDTs presented in chapter 2 showed that these FDTs still have considerable weaknesses concerning their architectural semantics in the area of communication systems, their level of abstraction, or their ease of use. For some of these FDTs tool development is a complex task and sometimes the realized tools don't fulfill the purpose they have been intended for. Not all FDTs support a constructive development approach. Since the formalization step demands considerable effort, FDTs should support a constructive approach and allow to develop tools which extensively utilize the formal descriptions.

The rule-based formal description technique CRS has been adopted to serve the existing needs. Employing the rule paradigm aimed at providing appropriate means to achieve intuitive and intelligible formal protocol specifications in relatively short time. Concerning the development of tools, it could be demonstrated that rules are a general and flexible mechanism for representing the specification knowledge and that they can be used by tools supporting various protocol engineering activities.

In order to apply CRS to the specification of communication services and protocols, a number of modifications and extensions have been developed. By adding operations to the ISO language ASN.1 an appropriate data definition language was integrated into CRS. As a result of the experiences made with CRS when specifying a number of real protocols, the meaning of rule contexts was slightly changed and the view concept was added. An operational formal semantics has been defined for the resulting technique to establish the basis for developing tools and to relate it to other FDTs.

The interpretation of CRS specifications according to the operational semantics is the basis for a constructive approach to the development of protocols. Existing implementation concepts for rule-based systems and for data representation have been employed to realize a compiler tool and runtime interpreters for the simulated execution of CRS specifications. The predicate logic formulas used in the specifications have to comply with a certain form that allows static code generation.

By organizing the CRS interpretation mechanisms in such a way that they can be invoked and used for knowledge processing and retrieval, executable CRS specifications can be exploited by different application tools. The general architecture of an integrated tools environment has been presented that combines protocol engineering tools in an efficient and flexible way. Application tools may use different shells in an exclusive or interleaved fashion to achieve their goals. The shells abstract from the internal representation of the formal descriptions in the environment kernel and provide different abstract views on the specified system. Depending on the particular development activity to be supported, different views are appropriate.

6.2 Implementation Status and Early Experiences

The technique CRS in the form presented in this thesis (incorporating X-ASN.1 and using an extended set of structuring constructs) has been developed during several specification case studies. These case studies included a number of well-known small examples ("Readers/Writers Problem", "Dining Philosophers", "Alternating-Bit Protocol", "Two-Phase-Commit" etc.), but, more important, also several existing protocol standards (OSI session layer protocol, transport layer service, X.25 packet level protocol, and OSI network service). The specifications were produced by students who did not know CRS or X-ASN.1 before they started. Most notably, all the specifications could be completed in time frames ranging from several days to several months.

About half a year was required by two students to develop the complete X.25 packet level protocol specification, including X-ASN.1 type definitions for all X.25 packets and rules for call setup, connection handling, data transfer, fragmenting/reassembling and retransmission of packets, call clearing, and the restart procedures. The final specification achieved was discussed with X.25 experts. It turned out that the rules describing the protocol could easily be understood and that the specification reflected the standard very well. An interesting experience during the discussions was that the experts could sometimes give no answers to the questions arising out of those parts of the specification that were most difficult to develop. Most of the times this was because the standard is not precise or contains no information at all. In fact, the corresponding protocol functions often have never been implemented (for example, quality of service, retransmission).

The local runtime environment for CRS specifications has been implemented completely, including the generic interpreters for CRS rule systems and gates, the scheduler, and the X-ASN.1 structure and data tree management. First prototypes of the CRS compiler and of the X-ASN.1 preprocessor generating the specification-specific code for the runtime environment have been completed as well. Most of the specification examples mentioned above have been compiled into executable form. The transport service specification and a subset of the X.25 packet level protocol specification are operational. The restrictions imposed on the form of predicate logic formulas were found acceptable for the trial specifications, although some parts of the original specifications had to be adapted in order to allow the transformation into executable form. The modifications made are a matter of style, however, and do not change the behavior of the protocols.

One major problem encountered in the CRS compiler described in section 5.1 is how to determine at compile time whether access operations referring to components of structured variables deliver already defined or undefined values. A solution has been developed for record structures, but for dynamic arrays this can only be decided at runtime. Therefore, the implementation of an interpretation-based approach has been started in order to overcome the problem and to retain all the information contained in the specification.

A first version of the application program interface constituting the rule-based shell, as well as the state-based and event-based shells of the integrated tools environment have been completed. After the principal mode of invocation and the data structures exchanged at the interfaces had been fixed, the first set of shell functions could easily be implemented. Additional functions that became necessary in the following could easily be realized. The concept of 'constraint trees' as explained in section 5.2 is currently under implementation.

The implementation of the application tools for simulation, test derivation, and test validation in the integrated tools environment is in progress. The CRS simulator is already in use and several errors could be found in specifications, leading to corrections or even to a different design. Support for adding event requests and offers at external gates and for backtracking during the simulation of specifications still has to be added.

The executable X.25 packet level protocol specification serves as a testbed for the activities on test case development and validation. Some of the components of TTCN test suites (declarations, behavior trees) can already be generated from a CRS specification using the test case development tool. Some work has still to be done concerning the constraints part of TTCN test suites. It is intended to provide test derivation strategies and heuristics in order to support test case design.

Some of the event sequences prescribed by the X.25 packet level test suite have already been checked using the test case validation tool. So far, it has been detected that admissible alternative behavior is missing in some of the test cases. Support for checking data constraints also has to be added in the test case validation tool and some work remains to be done concerning the mapping of TTCN types to ASN.1. The experiences made so far show that the interworking of the tools in the integrated environment poses no difficulties; rather large parts of the system tools and of the kernel representation can be shared.

6.3 Prospects and Future Work

More results from the application of FDTs to the development of real protocols are required before a final judgement on the suitability of formal techniques can be given. The experiences gained so far showed that FDTs have to be provided together with a development methodology and powerful tools in order to be useful. If this can be achieved, the prospects are encouraging.

The area of telecommunication and networking is rapidly expanding and distributed applications become popular. The protocols that are required to effectively provide flexible communication services for these applications are changing. The concept of symmetric end-to-end communication between peer entities underlying most OSI protocols seems to be inadequate for many applications. Multi-peer protocols, client-server models, and more application-oriented services have to be developed. FDTs and tools will help to design these services and protocols, to understand and reason about the basic problems, and to implement and test the corresponding communication systems. Extensions to the existing FDT concepts may be necessary. In CRS, for example, multicast communication mechanisms, or object-oriented extensions, such as inheritance and class hierarchies, will be useful to support the design of distributed systems.

Many new applications are possible given the transmission rates of optical fibers and the installation of high-speed networks (FDDI, B-ISDN etc.). These networks will also require new protocols and will emphasize the importance of non-functional protocol aspects. New language constructs, corresponding semantics, and formal methods for specifying, simulating, and predicting the performance of communication systems have to be included into FDTs. The formal models can be used to identify bottlenecks, to validate performance measures, and to come up with better implementations. Extensions to FDTs have to be developed for specifying timing aspects. This does also apply to CRS.

Communication in open systems, the field where FDTs are increasingly applied, is the key to open distributed processing. A formal development methodology, FDTs, and corresponding tools can provide valuable support for the design of open distributed applications and systems. The questions of how to decompose into distributed components, how the components interact, what are the implementation choices etc. can be approached by applying formal techniques. The techniques can also help to separate issues at different abstraction levels, to manage design complexity, to automate development steps, and to support the implementation of distributed systems. Support of this kind is needed in order to build applications that make use of the technology already available today.

References

[Acet87] **L. Aceto, R. DeNicola, A. Fantechi** *Testing Equivalences for Event Structures* Mathematical Models for the Semantics of Parallelism, M. Venturini Zilli (editor), Lecture Notes in Computer Science 280, Springer 1987, pp. 1-20

[Agga88] **S. Aggarwal, D. Barbara, K.Z. Meth** *A Software Environment for the Specification and Analysis of Problems of Coordination and Concurrency* IEEE Transactions on Software Engineering, Vol.14, No.3, March 1988, pp. 281-290

[Alde90] **R. Alderden** *COOPER - The Compositional Construction of a Canonical Tester* Proc. 2nd Int. Conf. on Formal Description Techniques for Communication Protocols and Distributed Systems, S. Vuong (editor), North-Holland 1990 pp. 13-17

[Atle90] **M. Atlevi** *SDT - a Real-time CASE Tool for the CCITT Specification Language SDL* Proc. 2nd Int. Conf. on Formal Description Techniques for Communication Protocols and Distributed Systems, S. Vuong (editor), North-Holland 1990, pp. 37-41

[Aujl89] **S.S. Aujla, M. Fletcher** *The Boyer-Moore Theorem-Prover and LOTOS* Formal Description Techniques, Proc. 1st Int. Conf. on Formal Description Techniques for Communication Protocols and Distributed Systems, K.J. Turner (editor), North-Holland 1989, pp. 169-184

[Baer89] **U. Bär** *Werkzeuge zur Entwicklung und Ausführung von OSI Conformance Tests für TTCN* Masters Thesis, University of Kaiserslautern (Germany), 1989 (in German)

[Barr81] **A. Barr, E.A. Feigenbaum** *The Handbook of Artificial Intelligence* HeurisTech Press, W. Kaufmann, Inc. 1981

[Baum90] **B. Baumgarten, A. Giessler, R. Platten** *Test Derivation from Net Models* Proc. 2nd Int. Workshop on Protocol Test Systems, W. Effelsberg, L. Mackert, J. de Meer (editors), North-Holland 1990, pp. 141-159

[Beli89] **F. Belina, D. Hogrefe, S. Trigila** *Modelling OSI in SDL* Formal Description Techniques, Proc. 1st Int. Conf. on Formal Description Techniques for Communication Protocols and Distributed Systems, K.J. Turner (editor), North-Holland 1989, pp. 135-142

[Beve89] **M. Bever, M. Zimmermann** *Data Presentation in Heterogenous Computer Networks* Technical Report No. 43.8905, IBM European Networking Center Heidelberg, 1989

[Bill86] **J. Billington, M.C. Wilbur-Ham, M.Y. Bearman** *Automated Protocol Verification* Proc. 5th Int. Symp. on Protocol Specification, Testing and Verification, M.Diaz (editor), North-Holland 1986, pp. 59-69

[Bill88] **J. Billington, G.R.Wheeler, M.C. Wilbur-Ham** *PROTEAN: A High-level Petri Net Tool for the Specification and Verification of Communication Protocols* IEEE Transactions on Software Engineering, Vol.14, No.3, March 1988, pp. 301-316

[Blac89] **U.D. Black** *Data Networks* Prentice-Hall 1989

[Blum83] **T.P. Blumer, D.P. Sidhu** *Experience with an Automated Protocol Development System* Proc. 3rd Int. Symp. on Protocol Specification, Testing and Verification, H. Rudin, C.H. West (editors), North-Holland 1983, pp. 369-380

[Boch77] **G. v. Bochmann, J. Gecsei** *A unified Method for the Specification and Verification of Protocols* Proc. IFIP 77, 1977, pp. 229-234

[Boch78] **G. v. Bochmann** *Finite State Description of Communication Protocols* Computer Networks, No.2, 1978, pp. 361-372

[Boch80a] **G. v. Bochmann, C.A. Sunshine** *Formal Methods in Communication Protocol Design* IEEE Transactions on Communications, Vol.28, No.4, April 1980, pp. 624-631

[Boch80b] **G. v. Bochmann** *A General Transition Model for Protocols and Communication Services* IEEE Transactions on Communications, Vol.28, No.4, April 1980, pp. 643-650

[Boch86] **G. v. Bochmann, B. Sarikaya, W.L. de Souza, H. Ural** *Use of PROLOG for Building Protocol Design Tools* Proc. 5th Int. Symp. on Protocol Specification, Testing and Verification, M.Diaz (editor), North-Holland 1986, pp. 131-147

[Boch87a] **G. v. Bochmann** *Usage of Protocol Development Tools: the Results of a Survey* Proc. 7th Int. Symp. on Protocol Specification, Testing and Verification, H. Rudin, C.H. West (editors), North-Holland 1987, pp. 139-161

[Boch87b] **G. v. Bochmann, G.W. Gerber, J.-M. Serre** *Semiautomatic Implementation of Communication Protocols* IEEE Transactions on Software Engineering, Vol.13, No.9, September 1987, pp. 989-1000

[Boch90a] **G. v. Bochmann, M. Deslauriers** *Combining ASN.1 Support with the LOTOS Language* Proc. 9th Int. Symp. on Protocol Specification, Testing and Verification, E. Brinksma, G. Scollo, C.A. Vissers (editors), North-Holland 1990

[Boch90b] **G. v. Bochmann** *Protocol Specification for OSI* Computer Networks and ISDN Systems, Vol.18, No.3, 1990, pp. 167-184

[Bolo87] **T. Bolognesi, E. Brinksma** *Introduction to the ISO Specification Language LOTOS* Computer Networks and ISDN Systems, Vol.14, No.1, 1987, pp. 25-59

[Boot90] **R. Booth** *An Evaluation of the LCF Theorem Prover using LOTOS* Proc. 2nd Int. Conf. on Formal Description Techniques for Communication Protocols and Distributed Systems, S. Vuong (editor), North-Holland 1990, pp. 83-99

[Bria87] **J.P.Briand, M.C. Fehri, L. Logrippo, A. Obaid** *Executing LOTOS Specifications* Proc. 6th Int. Symp. on Protocol Specification, Testing and Verification, B. Sarikaya, G. v. Bochmann (editors), North-Holland 1987, pp. 73-84

[Brin87] **E. Brinksma, G. Scollo, C. Steenbergen** *LOTOS Specifications, their Implementations and their Tests* Proc. 6th Int. Symp. on Protocol Specification, Testing and Verification, B. Sarikaya, G. v. Bochmann (editors), North-Holland 1987, pp. 349-360

[Brin88a] **E. Brinksma** *On the Design of Extended LOTOS* Doctoral Dissertation, University of Twente (The Netherlands), 1988

[Brin88b] **E. Brinksma** *A Theory for the Derivation of Tests* Proc. 8th Int. Symp. on Protocol Specification, Testing and Verification, S. Aggarwal, K. Sabnani (editors), North-Holland 1988, pp. 63-74

[Brin90] **E. Brinksma, R. Alderen, R. Langerak, J. v.d. Lagemaat, J. Tretmans** *A Formal Approach to Conformance Testing* Proc. 2nd Int. Workshop on Protocol Test Systems, W. Effelsberg, L. Mackert, J. de Meer (editors), North-Holland 1990, pp. 349-363

[Broy89] **M. Broy (editor)** *Constructive Methods in Computing Science* Proc. of the NATO Advanced Study Institute on Constructive Methods in Computing Science, Springer 1989

[Budk87] **S. Budkowski, P. Dembinski** *An Introduction to Estelle: A Specification Language for Distributed Systems* Computer Networks and ISDN Systems, Vol.14 (1), 1987, pp. 3-23

[Burg90] **S.P. v.d. Burgt, J. Kroon, E. Kwast, H.J. Wilts** *The RNL Conformance Kit* Proc. 2nd Int. Workshop on Protocol Test Systems, W. Effelsberg, L. Mackert, J. de Meer (editors), North-Holland 1990, pp. 279-294

[Card85] **L. Cardelli, P. Wegner** *On Understanding Types, Data Abstraction, and Polymorphism* Computing Surveys, Vol.17, No.4, December 1985, pp. 471-522

[CCZ100] **Committee Consultativ International Telegraphique et Telephonique** *Specification and Description Language SDL* CCITT SG X Recommendation Z.100, Com X-R15-E, 1987

[Chan90] **R.I. Chan, B.R. Smith, G. Neufeld, S.T. Chanson, W.B. Davis, S.T. Vuong, H.L. See, S. Chan** *A Software Environment for OSI Protocol Testing Systems* Proc. 9th Int. Symp. on Protocol Specification, Testing and Verification, E. Brinksma, G. Scollo, C.A. Vissers (editors), North-Holland 1990

[Chap89] **D. Chappell** *Abstract Syntax Notation One (ASN.1)* Journal of Data and Computer Communications, Vol. 1, No. 4, 1989, pp. 36-46

[Chen90] **K.E. Chen, L.N. Jackson** *Automatic Translation of SDL Specifications to Implementation based on Syntactic Transformation* Proc. 2nd Int. Conf. on Formal Description Techniques for Communication Protocols and Distributed Systems, S. Vuong (editor), North-Holland 1990, pp. 409-419

[Choq86] **N. Choquet, L. Fribourg, A. Mauboussin** *Runnable Protocol Specifications Using the Logic Interpreter SLOG* Proc. 5th Int. Symp. on Protocol Specification, Testing and Verification, M.Diaz (editor), North-Holland 1986, pp. 149-168

[Chow88] **C.-H. Chow, S.S. Lam** *PROSPEC: An Interactive Programming Environment for Designing and Verifying Communication Protocols* IEEE Transactions on Software Engineering, Vol.14, No.3, March 1988, pp. 327-338

[Chun88] **A. Chung, D.P. Sidhu, T.P. Blumer** *Automated Validation of Protocols Using EDS* Proc. 8th Int. Symp. on Protocol Specification, Testing and Verification, S. Aggarwal, K. Sabnani (editors), North-Holland 1988, pp. 351-360

[Clar89] **F. Clarius** *Entwurf und Implementierung einer rechnerinternen Repräsentationsform für eine regelbasierte Spezifikationstechnik* Masters Thesis, University of Kaiserslautern (Germany), 1989 (in German)

[Clea86] **J.C. Cleaveland** *An Introduction to Data Types Addision-Wesley 1986*

[Cleg89] **C. Cleghorn, H. Ural** *ASNST: an Abstract Syntax Notation One Support Tool* Computer Communications, Vol. 12, No. 5, October 1989, pp. 259-265

[Cloc81] **W.F. Clocksin, C.S. Mellish** *Programming in Prolog* Springer 1981

[Cour87] **J.P. Courtiat** *How Could Estelle Become a Better FDT ?* Proc. 7th Int. Symp. on Protocol Specification, Testing and Verification, H. Rudin, C.H. West (editors), North-Holland 1987, pp. 43-60

[Dant80] **A.A.S. Danthine** *Protocol Representation with Finite-State Models* IEEE Transactions on Communications, Vol.28, No.4, April 1980, pp. 633-641

[DeNi87] **R. De Nicola** *Extensional Equivalences for Transition Systems* Acta Informatica, Vol. 24, 1987, pp. 211-237

[Dhas86] **C.R. Dhas, V.K. Konangi** *X.25: an Interface to Public Packet Networks* IEEE Communications, September 1986, pp. 18-25

[Diaz86] **M. Diaz** *Petri Net Based Models in the Specification and Verification of Protocols* Petri Nets: Central Models and Their Properties, W.Brauer, W. Reisig, W. Rozenberg (editors), Lecture Notes in Computer Science 254, Springer 1987, pp. 135-170

[Ding90] **J. Dinger, K. Rothermel, K. Urbschat** *ECSE: An Efficient Environment for Layered Communication Protocols* Proc. 2nd IEEE Workshop on Future Trends of Distributed Computing Systems, Cairo (Egypt), 1990, pp. 305-314

[Drob89] **O. Drobnik** *Spezifikation, Verifikation und Testen von Protokollen* Tutorial, GI/ITG Fachtagung, University of Stuttgart (Germany), 1989 (in German)

[Duke88] **R. Duke, I. Hayes, P. King, G. Rose** *Protocol Specification and Verification Using Z* Proc. 8th Int. Symp. on Protocol Specification, Testing and Verification, S. Aggarwal, K. Sabnani (editors), North-Holland 1988, pp. 33-46

[Duke90] **R. Duke, G. Rose, A. Lee** *Object-oriented Protocol Specification* Proc. 10th Int. Symp. on Protocol Specification, Testing and Verification, L. Logrippo, R.L. Probert, H. Ural (editors), North-Holland (to be published)

[Dwye90] **D.J. Dwyer** *Progress in Conformance Testing Services ?* Proc. 2nd Int. Workshop on Protocol Test Systems, W. Effelsberg, L. Mackert, J. de Meer (editors), North-Holland 1990, pp. 17-27

[Effe86] **W. Effelsberg, A. Fleischmann** *Das ISO-Referenzmodell für offene Systeme und seine sieben Schichten* Informatik-Spektrum, Vol.9, 1986, pp. 280-299 (in German)

[Ehri85] **H. Ehrig, B. Mahr** *Fundamentals of Algebraic Specification 1* Springer 1985

[Eijk88] **P. v. Eijk** *Software Tools for the Specification Language LOTOS* Doctoral Dissertation, University of Twente (The Netherlands), 1988

[Eijk89a] **P.H.J. van Eijk, C.A. Vissers, M. Diaz** *The Formal Description Technique LOTOS* North-Holland 1989

[Eijk89b] **P.H.J. v. Eijk** *The Design of a Simulator Tool* The Formal Description Technique LOTOS, P.H.J. v. Eijk, C.A. Vissers, M. Diaz (editors), North-Holland 1989, pp. 351-390

[Eijk90a] **P. v. Eijk** *Tools for LOTOS Specification Style Transformation* Proc. 2nd Int. Conf. on Formal Description Techniques for Communication Protocols and Distributed Systems, S. Vuong (editor), North-Holland 1990, pp. 43-51

[Eijk90b] **P. v. Eijk, H. Kremer, M. v. Sinderen** *On the Use of Specification Styles for Automated Protocol Implementation from LOTOS to C* Proc. 10th Int. Symp. on Protocol Specification, Testing and Verification, L. Logrippo, R.L. Probert, H. Ural (editors), North-Holland (to be published)

[Ensl78] **P.H. Enslow** *What is a "Distributed" Data Processsing System ?* Computer, Vol.11, 1978, pp. 13-21

[Favr87] **J.-P. Favreau, R.J. Linn** *Automatic Generation of Test Scenario Skeletons from Protocol Specifications written in ESTELLE* Proc. 6th Int. Symp. on Protocol Specification, Testing and Verification, B. Sarikaya, G. v. Bochmann (editors), North-Holland 1987, pp. 191-202

[Feld87] **F. Feldbrugge, K. Jensen** *Petri Net Tool Overview* Petri Nets: Central Models and Their Properties, W.Brauer, W. Reisig, W. Rozenberg (editors), Lecture Notes in Computer Science 254, Springer 1987, pp. 20-61

[Flei87a] **A. Fleischmann, S.T. Chin, W. Effelsberg** *Specification and Implementation of an ISO Session Layer* Technical Report No. 43.8706, IBM European Networking Center, Heidelberg 1987

[Flei87b] **A. Fleischmann** *PASS - The Parallel Activity Specification Scheme* Technical Report No. 43.8715, IBM European Networking Center, Heidelberg 1987

[Flei88] **A. Fleischmann, T. Kunz, S. Pappe, K. Urbschat** *A Toolkit for Specifying and Implementing Communication Protocols* Proc. IFIP Symposium on Network Information Processing Systems, K. Boyanov (editor), North-Holland 1988

[Flei89] **A. Fleischmann, M. Bever, T.Kunz, S. Pappe, K. Urbschat** *The Development of Communication Software using the PASS Technology* Proc. 9th Int. Conf. on Computer Communication, J. Raviv (editor), North-Holland 1989

[Fros86] **R.A. Frost** *Introduction to Knowledge Base Systems* Collins 1986

[Gilb87] **D. Gilbert** *Executable LOTOS: Using PARLOG to implement an FDT* Proc. 7th Int. Symp. on Protocol Specification, Testing and Verification, H. Rudin, C.H. West (editors), North-Holland 1987, pp. 281-294

[Gilb89] **D. Gilbert** *A LOTOS to PARLOG Translator* Formal Description Techniques, Proc. 1st Int. Conf. on Formal Description Techniques for Communication Protocols and Distributed Systems, K.J. Turner (editor), North-Holland 1989, pp. 31-44

[Gora87] **W. Gora, R. Speyerer** *Abstract Syntax Notation One* DATACOM, No. 4, 1987, pp. 78-85 (in German)

[Gora88a] **W. Gora, R. Speyerer** *Automatic Code Generation for ASN.1-based Protocols* Proc. Enterprise '88, Baltimore, Maryland

[Gora88b] **W. Gora, R. Speyerer** *Das aktuelle Schlagwort: ASN.1* Informatik-Spektrum, Vol. 11, 1988, pp. 207-211 (in German)

[Gora90] **W. Gora** *ASN.1 Stand und Trends* DATACOM, No. 2, 1990, pp. 114-122 (in German)

[Gotz89] **R. Gotzhein** *On the Formal Definition and Representation of Architectural Concepts* Report No. 142, University of Hamburg, FBI-HH-B-142/89, 1989

[Gotz90a] **R. Gotzhein** *The Formal Definition of the Architectural Concept 'Interaction Point'* Proc. 2nd Int. Conf. on Formal Description Techniques for Communication Protocols and Distributed Systems, S. Vuong (editor), North-Holland 1990, pp. 67-81

[Gotz90b] **R. Gotzhein** *Specifying Communication Services with Temporal Logic* Proc. 10th Int. Symp. on Protocol Specification, Testing and Verification, L. Logrippo, R.L. Probert, H. Ural (editors), North-Holland (to be published)

[Gras90] **W. Grassl, H. Kossmann** *Automated Implementation and Testing of SDL Protocol Specifications* Proc. 2nd Int. Workshop on Protocol Test Systems, W. Effelsberg, L. Mackert, J. de Meer (editors), North-Holland 1990, pp. 201-204

[Guil88] **R. Guillemot, M. Haj-Hussein, L. Logrippo** *Executing Large LOTOS Specifications* Proc. 8th Int. Symp. on Protocol Specification, Testing and Verification, S. Aggarwal, K. Sabnani (editors), North-Holland 1988, pp. 399-410

[Haye87] **I. Hayes** *Specification Case Studies* Prentice-Hall 1987

[Haye90] **I.J. Hayes, M. Mowbray, G.A. Rose** *Signalling System No. 7 - The Network Layer* Proc. 9th Int. Symp. on Protocol Specification, Testing and Verification, E. Brinksma, G. Scollo, C.A. Vissers (editors), North-Holland 1990

[Hoar78] **C.A.R. Hoare** *Communicating Sequential Processes* Communications of the ACM, Vol.21, No.8, August 1978, pp. 666-677

[Hoar85] **C.A.R. Hoare** *Communicating Sequential Processes* Prentice-Hall 1985

[Hogr89a] **D. Hogrefe** *Estelle, LOTOS und SDL - Standard-Spezifikationssprachen für verteilte Systeme* Springer 1989 (in German)

[Hogr89b] **D. Hogrefe** *Nondeterminism and SDL* Formal Description Techniques, Proc. 1st Int. Conf. on Formal Description Techniques for Communication Protocols and Distributed Systems, K.J. Turner (editor), North-Holland 1989, pp. 157-167

[Holz83] **G. Holzmann, R.A. Beukers** *The PANDORA Protocol Development System* Proc. 3rd Int. Symp. on Protocol Specification, Testing and Verification, H. Rudin, C.H. West (editors), North-Holland 1983, pp. 357-368

[Hopc79] **J.E. Hopcroft, J.D. Ullman** *Introduction to Automata Theory, Languages and Computation* Addison-Wesley 1979

[ISO7498] **International Organization for Standardization (ISO)** *Open Systems Interconnection - Basic Reference Model* IS 7498

[ISO8208] **International Organization for Standardization (ISO)** *X.25 Packet Level Protocol for Data Terminal Equipment* IS 8208

[ISO8348] **International Organization for Standardization (ISO)** *Network Service Definition* IS 8348

[ISO8807] **International Organization for Standardization (ISO)** *LOTOS - A Formal Description Technique Based On The Temporal Ordering of Observational Behaviour* IS 8807

[ISO8824] **International Organization for Standardization (ISO)** *Specification of Abstract Syntax Notation One (ASN.1)* IS 8824

[ISO8825] **International Organization for Standardization (ISO)** *Specification of Basic Encoding Rules for Abstract Syntax Notation One (ASN.1)* IS 8825

[ISO8878] **International Organization for Standardization (ISO)** *Use of X.25 to provide the OSI connection-mode Network Service* IS 8878

[ISO9074] **International Organization for Standardization (ISO)** *Estelle - a Formal Description Technique Based On An Extended State Transitions Model* IS 9074

[ISO9646] **International Organization for Standardization (ISO)** *OSI Conformance Testing Methodology and Framework* DIS 9646

[Jard88] **C. Jard, J.-F. Monin, R. Groz** *Development of Veda, a Prototyping Tool for Distributed Algorithms* IEEE Transactions on Software Engineering, Vol.14, No.3, March 1988, pp. 339-352

[Jens87] **K. Jensen** *Coloured Petri Nets* Petri Nets: Central Models and Their Properties, W.Brauer, W. Reisig, W. Rozenberg (editors), Lecture Notes in Computer Science 254, Springer 1987, pp. 248-299

[Joua87] **J.-P. Jouannaud, P. Lescanne** *Rewriting Systems* Technology and Science of Informatics, Vol.6, No.3, 1987, pp. 181-199

[Karj86] **G. Karjoth** *An Interactive System for the Analysis of Communicating Processes* Proc. 5th Int. Symp. on Protocol Specification, Testing and Verification, M.Diaz (editor), North-Holland 1986, pp. 91-102

[Karj88] **G. Karjoth** *Implementing Process Algebra Specifications by State Machines* Proc. 8th Int. Symp. on Protocol Specification, Testing and Verification, S. Aggarwal, K. Sabnani (editors), North-Holland 1988, pp. 47-60

[Kell76] **R.M. Keller** *Formal Verification of Parallel Programs* Communications of the ACM, Vol. 19, No. 7, July 1976, pp. 371-384

[Kort86] **H. Korth, A. Silberschatz** *Database System Concepts* McGraw-Hill 1986

[Kowa79] **R. Kowalski** *Algorithm = Logic + Control* Communications of the ACM, Vol. 22, No. 7, July 1979, pp. 424-436

[Kric88] **R. Krickhahn, M.-J. Schachter-Radig** *Grundkonzepte der regelorientierten Programmierung* Informationstechnik it, No.6, 1988, pp. 434-445 (in German)

[Krum84] **H. Krumm** *Spezifikation, Implementierung und Verifikation von Kommunikationsdiensten* Doctoral Dissertation, University of Karlsruhe (Germany), 1984 (in German)

[Linn90] **R.J. Linn** *Conformance Testing for OSI Protocols* Computer Networks and ISDN Systems, Vol.18, No.3, 1990, pp. 203-219

[Lisk86] **B. Liskov, J. Guttag** *Abstraction and Specification in Program Development* MIT Press McGraw-Hill 1986

[Loec87] **J. Loeckx, K. Sieber** *The Foundations of Program Verification* Wiley-Teubner 1987

[Logr88] **L. Logrippo, A. Obaid, J.P.Briand, M.C. Fehri** *An Interpreter for LOTOS, A Specification Language for Distributed Systems* Software-Practice and Experience, Vol.18(4), April 1988, pp. 365-385

[Mack87] **L.F. Mackert, I.B. Neumeier-Mackert** *Communicating Rule Systems* Proc. 7th Int. Symp. on Protocol Specification, Testing and Verification, H. Rudin, C.H. West (editors), North-Holland 1987, pp. 77-88

[Mack88] **L.F. Mackert, R. Meyer, U. Scheere, J.M. Schneider, R.J. Velthuys, J. Burmeister, J. de Meer, I. Schrör** *A Generalized Conformance Test Tool for Communication Protocols* Proc. Int. Conf. on Distributed Computing Systems, San Jose CA. 1988

[Mana89] **J.A. Manas, T. de Miguel** *From LOTOS to C* Formal Description Techniques, Proc. 1st Int. Conf. on Formal Description Techniques for Communication Protocols and Distributed Systems, K.J. Turner (editor), North-Holland 1989, pp. 79-84

[Mang90] **T. Mangold** *Formal Specification of the X.25 Packet Layer in CRS* D.E.S.S. Thesis, University of Nancy (France), 1990

[Mann74] **Z. Manna** *Mathematical Theory of Computation* McGraw-Hill 1974

[Mast89] **J.J. Masterson, K.P. Ishaq, A.T. Hockley** *An Approach to Providing Support Tools for Formal Specification* Formal Description Techniques, Proc. 1st Int. Conf. on Formal Description Techniques for Communication

Protocols and Distributed Systems, K.J. Turner (editor), North-Holland 1989, pp. 1-14

[Mill90] **D. Miller** *Parsing and Normalizing CRS Specifications* Internal Report, IBM European Networking Center Heidelberg, 1990

[Miln80] **R. Milner** *A Calculus of Communicating Systems* Lecture Notes in Computer Science 92, Springer 1980

[Miln89] **R. Milner** *Communication and Concurrency* Prentice Hall 1989

[Mizu88] **T. Mizuno, J. Munemori, F. Sato, T. Nakakawaji, K. Katsuyama** *COTTAGE: Systematic Method for the Development of Communication Software* Proc. 8th Int. Symp. on Protocol Specification, Testing and Verification, S. Aggarwal, K. Sabnani (editors), North-Holland 1988, pp. 269-280

[Monn89] **J. Monnery** *A Rule-based Executable Specification of the X.25 Network Layer* Masters Thesis, Ecole Nationale Supérieur de Télécommunications de Bretagne, 1989

[Muel88] **H. Müller** *Höhere Petri-Netze - ein Werkzeug für Modellierung und Analyse verteilter Systeme und paralleler Datenverarbeitungsprozesse* Informationstechnik it, No.2, 1988, pp. 110-117 (in German)

[Nehm85] **J. Nehmer** *Softwaretechnik für verteilte Systeme* Springer 1985

[Nehm87] **J. Nehmer, D. Haban, F. Mattern, D. Wybranietz, H.D. Rombach** *Key Concepts of the INCAS Multicomputer Project* IEEE Transactions on Software Engineering, Vol.13, No.8, August 1987, pp. 913-923

[Neuf90] **G. Neufeld** *The Design and Implementation of an ASN.1-C Compiler* IEEE Transactions on Software Engineering, Vol.16, No.10, October 1990, pp. 1209-1220

[Neum87] **I. B. Neumeier-Mackert, L.F. Mackert, A. Fleischmann** *A Knowledge-based Protocol Engineering Environment* Proc. 1st Int. Conf. on Networking in Open Systems, Washington, 1988

[Neum88] **I. B. Neumeier-Mackert** *Modellierung und Implementierung kommunizierender Systeme mit Hilfe von Regeln* Doctoral Dissertation, University of Erlangen-Nürnberg (Germany), 1988 (in German)

[Papp87] **G. Pappalardo** *Experiences with a Verification and Simulation Tool for Behavioral Languages* Proc. 7th Int. Symp. on Protocol Specification,

Testing and Verification, H. Rudin, C.H. West (editors), North-Holland 1987, pp. 251-264

[Paul85] **C. Paule, H. Eckert** *Das Netzsimulationssystem NESSY* Arbeitspapiere der GMD 181, Gesellschaft für Mathematik u. Datenverarbeitung, Darmstadt (Germany), 1985 (in German)

[Prob89] **R.L. Probert, H. Ural, M.W.A. Hornbeek** *A Comprehensive Software Environment for Developing Standardized Conformance Test Suites* Computer Networks and ISDN Systems, Vol.18, 1989/90, pp. 19-29

[Rayn87] **D. Rayner** *OSI Conformance Testing* Computer Networks and ISDN Systems, Vol.14 (1), 1987, pp. 79-98

[Roor90] **S.B. Roorda** *CRS Supported Development of TTCN Test Cases* Masters Thesis, University of Twente (The Netherlands), 1990

[Rudi88] **H. Rudin** *Protocol Engineering: A Critical Assessment* Proc. 8th Int. Symp. on Protocol Specification, Testing and Verification, S. Aggarwal, K. Sabnani (editors), North-Holland 1988, pp. 3-16

[Sabn87] **K.K. Sabnani, A.M. Lapone** *PAV - Protocol Analyzer and Verifier* Proc. 6th Int. Symp. on Protocol Specification, Testing and Verification, B. Sarikaya, G. v. Bochmann (editors), North-Holland 1987, pp. 29-34

[Saqu89] **P. de Saqui-Sannes, J.-P. Courtiat** *ESTIM: The Estelle Simulator Prototype of the ESPRIT-SEDOS Project* Formal Description Techniques, Proc. 1st Int. Conf. on Formal Description Techniques for Communication Protocols and Distributed Systems, K.J. Turner (editor), North-Holland 1989, pp. 15-29

[Sari88] **B. Sarikaya, Q. Gao** *Translation of Test Specifications in TTCN to LOTOS* Proc. 8th Int. Symp. on Protocol Specification, Testing and Verification, S. Aggarwal, K. Sabnani (editors), North-Holland 1988, pp.219-230

[Sari89] **B. Sarikaya** *Conformance Testing: Architectures and Test Sequences* Computer Networks and ISDN Systems, Vol.17, 1989, pp. 111-126

[Sari90] **B. Sarikaya, V. Koukoulidis, G. v. Bochmann** *Method of analysing extended finite-state machine specifications* Computer Communications, Vol.13, No.2, March 1990, pp. 83-92

[Sche89] **U. Scheere** *Integration von ASN.1 in "C"* Project Thesis, University of Karlsruhe (Germany), 1989 (in German)

[Sche90] **U. Scheere** *ASN.1 Definition of X.25 Packets and X.25 Encoding Rules* Internal Report, IBM European Networking Center Heidelberg, 1990

[Schn87] **J.M. Schneider** *Spezifikation und Implementierung von OSI Conformance Tests* Masters Thesis, University of Kaiserslautern (Germany), 1987 (in German)

[Schn88] **J.M. Schneider** *Communicating Rule Systems - eine regelbasierte Spezifikationsumgebung* Proc. Workshop "Sprachspezifische Programmierumgebungen", Technical University of Darmstadt (Germany), April 1988 (in German)

[Schn89a] **J.M. Schneider, I.B. Neumeier-Mackert, L.F. Mackert, R.J. Velthuys** *Executing Protocol Knowledge* Technical Report No. 43.8908, IBM European Networking Center Heidelberg, 1989

[Schn89b] **J.M. Schneider, I.B. Neumeier-Mackert, L.F. Mackert, R.J. Velthuys** *Executable Communicating Rule Systems* Proc. 2nd Int. Conf. on Formal Description Techniques for Communication Protocols and Distributed Systems, S.T. Vuong (editor), North-Holland 1990 pp. 421-436

[Schn90a] **J.M. Schneider, H. Müller, L.F. Mackert, G. Zörntlein, R.J. Velthuys** *Transition System Semantics for Communicating Rule Systems* Technical Report No. 43.9005, IBM European Networking Center, Heidelberg 1990

[Schn90b] **J.M. Schneider, L. F. Mackert, G. Zörntlein, R.J. Velthuys, U. Bär** *An Integrated Environment for Developing Communication Protocols* Computer Networks and ISDN Systems (to be published)

[Schn90c] **J.M. Schneider, U. Bär, T. Mangold, J. Monnery, R.J. Velthuys, G. Zörntlein** *Formal Description of X.25 Applying a Rule-based Approach* Proc. GI/ITG "7. Fachtagung Kommunikation in verteilten Systemen", Springer (to be published)

[Scho90] **B. Schott** *Synthese prozeduraler Programme aus regelbasierten Spezifikationen* Masters Thesis, University of Kaiserslautern (Germany), 1990 (in German)

[Schw82] **R.L. Schwartz, P.M. Melliar-Smith** *From State Machines to Temporal Logic: Specification Methods for Protocol Standards* IEEE Transactions on Communications, Vol.30, No.12, December 1982, pp. 2486-2496

[Shir88] **N. Shiratori, S. Noguchi** *An Intelligent User-friendly Support System for Protocol and Communication Software Development* Proc. 8th Int. Symp. on Protocol Specification, Testing and Verification, S. Aggarwal, K. Sabnani (editors), North-Holland 1988, pp. 257-268

[Sidh83] **D. Sidhu** *Protocol Verification via Executable Logic Specifications* Proc. 3rd Int. Symp. on Protocol Specification, Testing and Verification, H. Rudin, C.H. West (editors), North-Holland 1983, pp. 237-248

[Sidh89] **D. Sidhu, T.-K. Leung** *Formal Methods for Protocol Testing: a Detailed Study* IEEE Transactions on Software Engineering, Vol. 15, No. 4, April 1989, pp. 413-426

[Sidh90a] **D. Sidhu, A. Chung** *Experience with Formal Methods in Protocol Development* Proc. 2nd Int. Conf. on Formal Description Techniques for Communication Protocols and Distributed Systems, S. Vuong (editor), North-Holland 1990 pp. 437-453

[Sidh90b] **D.P. Sidhu, T.P. Blumer** *Semi-automatic Implementation of OSI Protocols* Computer Networks and ISDN Systems, Vol.18, No.3, 1990, pp. 221-238

[Sije89] **R. Sijelmassi, P. Gaudette** *An Object-Oriented Model for ESTELLE* Formal Description Techniques, Proc. 1st Int. Conf. on Formal Description Techniques for Communication Protocols and Distributed Systems, K.J. Turner (editor), North-Holland 1989, pp. 91-105

[Slom87] **M. Sloman, J. Kramer** *Distributed Systems and Computer Networks* Prentice-Hall 1987

[Spiv89] **J.M. Spivey** *The Z Notation* Prentice Hall 1989

[Stol90] **W. Stoll** *KATE - a Test System for OSI Protocols* Proc. 2nd Int. Workshop on Protocol Test Systems, W. Effelsberg, L. Mackert, J. de Meer (editors), North-Holland 1990, pp. 215-229

[Svob89] **L. Svobodova** *Implementing OSI Systems* IEEE Journal on Selected Areas in Communications, Vol.7, No.7, September 1989, pp. 1115-1130

[Tane88] **A.S. Tanenbaum** *Computer Networks* Prentice-Hall 1988

[Thom90] **M. Thomas** *From 1 Notation to Another One: An ACT-ONE Semantics for ASN.1* Proc. 2nd Int. Conf. on Formal Description Techniques for Communication Protocols and Distributed Systems, S. Vuong (editor), North-Holland 1990 pp. 517-531

[Tret89] **J. Tretmans** *HIPPO: A LOTOS Simulator* The Formal Description Technique LOTOS, P.H.J. v. Eijk, C.A. Vissers, M. Diaz (editors), North-Holland 1989, pp. 391-396

[Tret90] **J. Tretmans** *Test Case Derivation from LOTOS Specifications* Proc. 2nd Int. Conf. on Formal Description Techniques for Communication Protocols and Distributed Systems, S. Vuong (editor), North-Holland 1990 pp. 345-359

[Turn87] **K.J. Turner** *An Architectural Semantics for LOTOS* Proc. 7th Int. Symp. on Protocol Specification, Testing and Verification, H. Rudin, C.H. West (editors), North-Holland 1987, pp. 15-28

[Turn90] **K. Turner** *A LOTOS-based Development Strategy* Proc. 2nd Int. Conf. on Formal Description Techniques for Communication Protocols and Distributed Systems, S. Vuong (editor), North-Holland 1990 pp. 117-132

[Ural87] **H. Ural** *A Test Derivation Method for Protocol Conformance Testing* Proc. 7th Int. Symp. on Protocol Specification, Testing and Verification, H. Rudin, C.H. West (editors), North-Holland 1987, pp. 347-358

[Velt89] **R.J. Velthuys, J.M. Schneider, L.F. Mackert** *Protocol Conformance Testing with Communicating Rule Systems* Proc. 2nd Int. Workshop on Protocol Test Systems, W. Effelsberg, L. Mackert, J. de Meer (editors), North-Holland 1990, pp. 231-252

[Velt90] **R.J. Velthuys, L.F. Mackert, J.M. Schneider, G. Zörntlein** *Structuring Mechanisms for the Formal Description Technique CRS* Technical Report, IBM European Networking Center Heidelberg (to be published)

[Viss86a] **C. Vissers, L. Logrippo** *The Importance of the Service Concept in the Design of Data Communications Protocols* Proc. 5th Int. Symp. on Protocol Specification, Testing and Verification, M.Diaz (editor), North-Holland 1986, pp. 3-17

[Viss86b] **C. Vissers, G. Scollo** *Formal Specification in OSI* Proc. Int. Seminar on Networking in Open Systems, Oberlech, Austria, August 1986, Lecture Notes in Computer Science 248, Springer 1987, pp 338-359

[Viss88] **C.A. Vissers, G. Scollo, M. v. Sinderen** *Architecture and Specification Style in Formal Descriptions of Distributed Systems* Proc. 8th Int. Symp. on Protocol Specification, Testing and Verification, S. Aggarwal, K. Sabnani (editors), North-Holland 1988, pp. 189-204

[Vogl89] **W. Vogler** *Failure Semantics and Deadlocking of Modular Petri Nets* Acta Informatica, Vol. 26, 1989, pp. 333-348

[Vuon87] **S.T. Vuong, D.D. Hui, D.D. Cowan** *VALIRA - a Tool for Protocol Validation via Reachability Analysis* Proc. 6th Int. Symp. on Protocol Specification, Testing and Verification, B. Sarikaya, G. v. Bochmann (editors), North-Holland 1987, pp. 35-41

[Vuon88] **S.T. Vuong, A.C. Lau, R.I. Chan** *Semiautomatic Implementation of Protocols Using an Estelle-C Compiler* IEEE Transactions on Software Engineering, Vol. 14, No.3, March 1988, pp. 384-393

[West87] **C.H. West** *Protocol Validation by Random State Exploration* Proc. 6th Int. Symp. on Protocol Specification, Testing and Verification, B. Sarikaya, G. v. Bochmann (editors), North-Holland 1987, pp. 233-242

[Weze90] **C. Wezeman** *The CO-OP Method for Compositional Derivation of Canonical Testers* Proc. 9th Int. Symp. on Protocol Specification, Testing and Verification, E. Brinksma, G. Scollo, C.A. Vissers (editors), North-Holland 1990

[Whee85] **G.R. Wheeler** *Numerical Petri Nets - A definition* Telecom Australia, Res. Labs. Rep. 7780, May 1985

[Wile89] **A. Wiles, B. Pehrson** *ITEX - An Interactive TTCN Editor and Executor* Swedish Institute of Computer Science, Spanga 1989

[Wins84] **P.H. Winston** *Artificial Intelligence* Addison-Wesley 1984

[Yemi83] **Y Yemini, N Nounou** *CUPID: A Protocol Development Environment* Proc. 3rd Int. Symp. on Protocol Specification, Testing and Verification, H. Rudin, C.H. West (editors), North-Holland 1983, pp. 347-355

Index

Z